Disposal of Waste Resulting From Arsenic Removal Processes

The mission of the Awwa Research Foundation (AwwaRF) is to advance the science of water to improve the quality of life. Funded primarily through annual subscription payments from over 1,000 utilities, consulting firms, and manufacturers in North America and abroad, AwwaRF sponsors research on all aspects of drinking water, including supply and resources, treatment, monitoring and analysis, distribution, management, and health effects.

From its headquarters in Denver, Colorado, the AwwaRF staff directs and supports the efforts of over 500 volunteers, who are the heart of the research program. These volunteers, serving on various boards and committees, use their expertise to select and monitor research studies to benefit the entire drinking water community.

Research findings are disseminated through a number of technology transfer activities, including research reports, conferences, videotape summaries, and periodicals.

Disposal of Waste Resulting From Arsenic Removal Processes

Prepared by:
David A. Cornwell, Michael J. MacPhee, and **Rodney Mutter**
Environmental Engineering & Technology, Inc.
712 Gum Rock Court, Newport News, VA 23606
and
John Novak and **Marc Edwards**
Virginia Polytechnic Institute and State University, Department of Civil Engineering
312 North Hall, Blacksburg, Virginia 24061

Jointly sponsored by:
Awwa Research Foundation
6666 West Quincy Avenue, Denver, CO 80235-3098
and
U.S. Environmental Protection Agency
Washington D.C.
and
Association of California Water Agencies
Sacramento, CA

Published by:
Awwa Research Foundation
Advancing the Science of Water®
IWA Publishing

DISCLAIMER

This study was jointly funded by the Awwa Research Foundation (AwwaRF), the U.S. Environmental Protection Agency (USEPA), and the Association of California Water Agencies (ACWA) under Cooperative Agreement No. CR 827268-01-0. AwwaRF, USEPA, and ACWA assume no responsibility for the content of the research study reported in this publication or for the opinions or statements of fact expressed in the report. The mention of trade names for commercial products does not represent or imply the approval or endorsement of AwwaRF, USEPA, or ACWA. This report is presented solely for informational purposes.

Published by IWA Publishing, Alliance House, 12 Caxton Street, London SW1H 0QS, UK

Telephone: +44 (0) 20 7654 5500; Fax: +44 (0) 20 7654 5555; Email: publications@iwap.co.uk; Web: **www.iwapublishing.com**

AwwaRF report number 90953F. Originally published by AwwaRF for its subscribers in 2003. IWA Publishing version published 2004.

ISBN 1 84339 855 9

CONTENTS

TABLES

FIGURES

FOREWORD

The Awwa Research Foundation is a nonprofit corporation that is dedicated to the implementation of a research effort to help utilities respond to regulatory requirements and traditional high-priority concerns of the industry. The research agenda is developed through a process of consultation with subscribers and drinking water professionals. Under the umbrella of a Strategic Research Plan, the Research Advisory council prioritizes the suggested projects based upon current and future needs, applicability, and past work; the recommendations are forwarded to the Board of Trustees for final selection. The foundation also sponsors research projects through the unsolicited proposal process; the Collaborative Research, Research Applications, and Tailored Collaboration programs; and various joint research efforts with organizations such as the U.S. Environmental Protection Agency, the U.S. Bureau of Reclamation, and the Association of California Water Agencies.

This publication is a result of one of these sponsored studies, and it is hoped that its findings will be applied in communities throughout the world. The following report serves not only as a means of communicating the results of the water industry's centralized research program but also as a tool to enlist the further support of the nonmember utilities and individuals.

Projects are managed closely from their inception to the final report by the foundation's staff and large cadre of volunteers who willingly contribute their time and expertise. The foundation serves a planning and management function and awards contracts to other institutions such as water utilities, universities, and engineering firms. The funding for this research effort comes primarily from the Subscription Program, through which water utilities subscribe to the research program and make an annual payment proportionate to the volume of water they deliver and consultants and manufacturers subscribe based on their annual billings. The program offers a cost-effective and fair method for funding research in the public interest. A broad spectrum of water supply issues is addressed by the foundation's research agenda; resources; treatment and operations; distribution and storage, water quality and analysis, toxicology, economics, and management. The ultimate purpose of the coordinated effort is to assist water suppliers to provide the highest possible quality of water economically and reliably. The true benefits are realized when the results are implemented at the utility level. The foundation's trustees are pleased to offer this publication as a contribution toward that end.

The reduced drinking water arsenic MCL of 10 $\mu g/L$ will require many utilities to remove more arsenic from their source water, thereby concentrating it in their residuals. Because arsenic removal is sensitive to both the pH and the oxidation state, any process that changes pH or results in a reducing environment may release arsenic from the solid phase. Management of arsenic-laden residuals is an important issue due to the potential re-release of arsenic into the environment. This report identifies factors that influence arsenic release and provides a decision tree for use by utilities with arsenic-laden residuals.

Edmund G. Archuleta, P.E.	James F. Manwaring, P.E.
Chair, Board of Trustees	Executive Director
Awwa Research Foundation	Awwa Research Foundation

ACKNOWLEDGMENTS

Environmental Engineering & Technology, Inc. (EE&T) and Virginia Tech would like to thank all of the utilities who participated in this research effort, as well as the dedicated Project Advisory Committee (PAC) who provided invaluable insight throughout completion of the project. The research team would like to acknowledge the research efforts provided by Virginia Tech graduate students Cortney Itle and Jeff Parks, who performed much of the laboratory research during the project.

The participating utilities assisting with the project included Louisiana Water Company, City of Great Falls, MT, City of Helena, MT, Billings (MT) Department of Public Utilities, Lockwood (MT) Water Users Association, City of Los Angeles Department of Water and Power, and Indiana-American Water Company.

The PAC members included Thomas Sorg, USEPA, Joseph Chwirka, CH2MHILL, and Robert Martin, East Valley Water District. The research team and PAC would also like to thank our AwwaRF Project Officers Emerson Lomaquahu and Alice Fulmer for their efforts supporting and managing this project.

EXECUTIVE SUMMARY

OVERVIEW

The management of arsenic (As)-laden residuals is an important issue due to the potential re-release of arsenic into the environment as a result of the various handling and disposal techniques commonly used by utilities. Because arsenic removal is sensitive to both the pH and the oxidation state, any process that changes pH or results in a reducing environment may release arsenic from the solid phase. These processes include chemical conditioning during dewatering, storage and lagooning, and ultimate disposal options such as landfilling, land application, and indirect discharge to a sanitary sewer. With the arsenic maximum contaminant level (MCL) of 10 μg/L modifications of the treatment processes are needed to meet these lower levels. Residuals handling and disposal will be further impacted and there are many questions regarding differences in environmental conditions and the retention of arsenic in residuals.

The objective of this research project was to better understand the factors that may cause the release of arsenic from solid residuals and, thereby, allow arsenic to re-enter the environment. The project tasks were structured in order to determine arsenic release from residuals due to chemical conditioning for dewatering, thickening, short- and long-term storage, as well as to determine innovative methods for preventing arsenic mobility from residuals into the environment. These research efforts will help provide the data necessary to develop recommendations for regulators and utilities to minimize arsenic release during handling, storage, and disposal of water treatment plant (WTP) residuals. The specific project tasks performed included:

- Residuals regulatory survey (federal and state regulations)
- Residuals chemical and physical characterization
- Laboratory treatability investigations
- Identification of factors that influence arsenic release
- Development of guidelines for utilities with arsenic-laden residuals

Characterization and laboratory treatability work were used to develop a database comparing paired toxicity characteristic leaching procedures (TCLP) and California Waste Extraction Test (Ca WET) results for the same residuals. These residuals were collected from full-scale coagulation, lime softening, and iron/manganese removal processes.

REGULATORY SURVEY

The purpose of the regulatory review was to work with federal and state agencies to compile and interpret the existing federal and state regulations that apply to the handling, transportation and disposal of arsenic-laden residuals. The regulatory review provided in this document is intended to be a continuation of the regulatory summary generated by the EPA in the report titled "Regulations on the Disposal of Arsenic Residuals" (USEPA 2000). The EPA report summarized the regulations from seven U.S. states: CA, AZ, NM, NV, ME, PA, and NE. This report provides state regulations from 10 additional U.S. states: MN, KS, LA, CO, IN, WI, UT, MT, NH and TX (see Figure ES.1).

Federal Regulatory Review

Although there are no federal regulations that specifically deal with the disposal of water treatment plant residuals, a number of federal regulations are in place that affect specific methods of disposal of these materials. In general, these regulations are meant to protect either the receiving water, land, or underlying groundwater at the location where residuals disposal occurs. This document provides a detailed overview of these regulations.

Residuals from water treatment plants are disposed of according to the physical form of the residual; that is, if the residual is a liquid or a solid. Liquid residuals can be disposed of by the following means:

- Direct discharge (National Pollutant Discharge Elimination System (NPDES) permit limits)
- Indirect discharge (Publicly Owned Treatment Works (POTW) pre-treatment limits)
- Underground injection (Underground Injection Control (UIC) Program)
- Beneficial use (Resource Conservation and Recovery Act (RCRA))

Solids residuals are typically disposed of by the following methods:

- Landfilling (RCRA)
- Beneficial use (RCRA)

Lagoons (including evaporation ponds in arid regions of the western U.S.) can also be used to manage liquid or solid residuals, however, these are not considered to be a final disposal option, only temporary handling. Each of these disposal methods is regulated by different requirements and regulations as outlined in this report. A detailed description of these federal and state programs for residuals disposal is provided in Chapter 2 of this report.

State Regulatory Review

The state regulatory survey conducted during this research project included a total of ten states as previously listed. The regulatory survey focused on disposal of both liquid and solid phase residuals. Disposal of liquid residuals could include direct discharge to surface water by NPDES permit, indirect discharge to a POTW, beneficial reuse, or underground injection. Solid residuals disposal alternatives could include a solid waste landfill, monofill, or a type of beneficial re-use such as land application or composting. Many utilities have lagoons or stockpiled residuals as a means of storage prior to final disposal.

Overall, the approach to residuals disposal regulations was similar for these ten states. For example, all states require an NPDES permit for direct stream discharge, and typically limit arsenic to 0.05 mg/L or lower. Indirect discharge (sanitary sewer) arsenic limits are based on state or EPA industrial pretreatment programs, with local or regional limits typically falling in the 0.05 to 1 mg/L range. Landfilling requires that residuals have no free liquids and be non-hazardous based on the TCLP test (arsenic less than 5.0 mg/L). In California, arsenic must be less than 5.0 mg/L as determined in the Ca WET test, and total arsenic must be less than 500 mg/kg. Some states have adopted the federal 503 biosolids metals limits for residuals reuse and require residuals to be non-hazardous based on the TCLP test.

CHEMICAL CHARACTERIZATION

The utilities sampled during the project are summarized in Table ES.1. Eight settled residuals samples and six spent filter backwash (SFBW) samples were collected. One of the lime softening residuals was collected near the end of the study, and consequently was used only for characterization.

Table ES.1

Sample collection locations

Utility name	Sample type	Chemical treatment
Louisiana Water Company (New Iberia), LA	Settled solids/SFBW	Lime softening
City of Great Falls, MT	Settled solids/SFBW	Alum
City of Helena, MT	Settled solids/SFBW	Alum
Dept. of Public Utilities, Billings, MT	Settled solids/SFBW	Ferric chloride
Lockwood Water Users Association, Lockwood, MT	Settled solids/SFBW	Ferric chloride
Los Angeles Dept. of Water & Power (LADWP), CA	Settled solids/SFBW	Ferric chloride
Indiana-American Water Company, IN	Settled solids	Ferric chloride
Heath, OH	Settled solids	Lime softening

Settled Solids Characterization

Each sample was divided into three portions in order to generate the different residuals streams required for analysis. Settled residuals collected from sedimentation basins (or from a lagoon in the case of LADWP's direct filtration facility) were completely mixed as delivered, and were referred to as untreated residuals. Secondary residuals were created from the separation of the liquid and solid portions. These residuals were gravity settled to separate the solids from the liquid portion and then the solids were either air- or oven-dried prior to analysis. The results from the chemical characterization analyses are summarized in Table ES.2 and analytical methodology is described in Chapter 4.

Table ES.2

Settled residuals characterization data from participating utilities

Utility name	Sample description	Solids conc. (percent)	pH	Total As*	Total Fe*	Total Al*	TCLP As (mg/L)	WET As (mg/L)
Louisiana Water Company	Untreated residuals	4.1	11.5	—	—	—	—	—
(New Iberia), LA	Liquid portion	—	11.3	BDL	460	610	—	—
	Solid portion	100.0	---	110	2,320	11,800	0.012	0.096
City of Great Falls, MT	Untreated residuals	3.0	7.5	—	—	—	—	—
	Liquid portion	—	7.3	280	290	430	—	—
	Solid portion	100.0	—	490	14,660	109,160	0.158	4.830
City of Helena, MT	Untreated residuals	3.6	6.7	—	—	—	—	—
	Liquid portion	—	7.7	31	3,720	210	—	—
	Solid portion	100.0	---	420	43,730	41,000	0.125	1.880
Dept. Of Public Utilities	Untreated residuals†	4.1	6.9	—	—	—	—	—
(Billings), MT	Liquid portion	---	7.6	2	1,570	440	—	—
	Solid portion	100.0	---	70	74,670	25,520	0.013	0.863
Lockwood Water Users	Untreated residuals	4.7	7.2	—	—	—	—	—
Association (Lockwood), MT	Liquid portion	—	7.4	BDL	26,660	400	—	—
	Solid portion	100.0	---	820	327,700	11,800	0.011	0.224
Los Angeles Dept. Of Water &	Untreated residuals	14.0	7.2	—	---	—	—	—
Power (LADWP), CA	Liquid portion	—	6.8	358	22,010	310	—	—
	Solid portion	100.0	---	730	139,390	16,990	0.162	4.931
Indiana-American Water	Untreated residuals	1.9	7.2	—	—	—	—	—
Company, IN	Liquid portion	—	7.7	14	500	210	—	—
	Solid portion	100.0	---	15,730	493,200	1,720	0.031	2.471
Heath, OH	Untreated residuals	26	10	—	—	—	—	—
	Liquid portion	—	9.4	BDL	BDL	2.0	—	—
	Solid portion	100.0	---	19	3,300	9,107	BDL	0.697

BDL= Below Detection Level (<2 μg/L for Arsenic)

*Units for mixed and liquid samples are μg/L; units for solid samples and mg/kg

†Completely mixed residuals before separation

The two highest arsenic concentrations in the liquid decant after gravity settling were from Los Angeles Department of Water and Power (LADWP) and Great Falls, which were 358 μg/L and 280 μg/L, respectively. These arsenic levels would prohibit the use of stream discharge and may limit discharge to a sanitary sewer without additional treatment such as chemical addition, while the discharge of the supernatant from the other utilities would not be restricted due to arsenic.

The arsenic concentration in the dried solids ranged from 19 mg/kg to 15,730 mg/kg. Only the Billings, MT (70 mg/kg) and the Heath, OH (19 mg/kg) solids could likely be used for land application or other beneficial use, depending on the state regulatory requirements. If the state has adopted the 40 CFR 503 biosolids land application limits then the ceiling limit for arsenic is 75 mg/kg.

All of the solids samples passed both the TCLP and Ca WET tests with supernatant arsenic levels less than 5.0 mg/L. The LADWP solids sample, however, was only slightly below the arsenic limit using the Ca WET test (see Figure ES.2). Based on these one-time analyses, the samples evaluated would be considered non-hazardous. Since the test is subject to technique, any Ca WET level greater than 1 mg/L may be of concern. These data also indicate significant differences between the TCLP arsenic and Ca WET arsenic concentrations measured from the solid residuals samples. These data indicated (using 100 percent dry solids) that the Ca WET extraction increased the leached arsenic concentration from eight to 80 times higher than the TCLP arsenic extraction. This increase is expected due to the use of citric acid buffer, 48-hour extraction period, and anaerobic test conditions for the Ca WET extraction, which is more aggressive than acetic acid extraction for TCLP analysis.

SFBW Residuals Characterization

SFBW samples from six different utilities were analyzed for total arsenic, total iron, total aluminum, TSS, pH, and turbidity. Results from these tests are presented in Table ES.3.

These data show that three of the six samples had arsenic concentrations greater than 50 μg/L and without solids removal, therefore, would exceed the NPDES limit for direct discharge. Depending on the local Publicly Owned Treatment Works (POTW) the Technically Based Local Limit (TBLL) established for arsenic, these residuals may or may not be acceptable for indirect discharge. The TBLLs for arsenic used by POTWs generally range from 50 to 1,000 μg/L.

Table ES.3

Results of SFBW characterization analysis

Utility name	Type	Turbidity (ntu)	pH	Total As (μg/L)	Total Fe (mg/L)	Total Al (mg/L)	Ca (mg/L)	Mg (mg/L)	TSS (mg/L)
Louisiana Water Co. (New Iberia), LA	Lime	1,150	8.6	43	18.6	1.9	267	18	2,364
City of Great Falls, MT	Alum	7.8	7.5	32	0.1	4.8	---	---	269
City of Helena, MT	Alum	265	7.1	163	9.2	21.9	---	---	515
Billings Dept. of Public Utilities (Billings), MT	Ferric	28	7.7	63	34.6	1.2	---	---	119
Lockwood Water Users Assoc. (Lockwood), MT	Ferric	12	7.3	14	8.7	0.4	---	---	28
Los Angeles Dept. Of Water & Power (LADWP), CA	Ferric	208	7.2	185	66.3	5.6	---	---	

LABORATORY TREATABILITY

The laboratory treatability research tasks conducted during this project included:

- Lagoon simulations
- Sand drying bed simulations
- SFBW clarification using ferric chloride and polymers
- Lime conditioning and dewatering simulations

These evaluations were performed to assess the factors influencing the release of arsenic from solid phase residuals.

Lagoon storage simulations were performed using settled residuals (most collected from sedimentation basins) from each of the utilities to determine under what conditions, and to what extent arsenic is released. Bench-scale sand drying bed simulations were used to determine the potential of arsenic to leach from sedimentation basin residuals into the underdrain collection system, or possibly into the groundwater. Bench-scale chemical precipitation tests, using both ferric chloride and polymer conditioning, were used to remove arsenic from SFBW samples. These bench-scale clarification tests used for treatment of SFBW during this project are commonly used in the water industry for simulating full-scale treatability. Finally, lime dewatering simulations were conducted to determine the release of arsenic from solid residuals when lime is used to condition residuals for dewatering.

Lagoon Simulations

In lagoon simulations, sample sludges were placed in a controlled environment in closed, air-tight containers, similar to conditions that might be expected in thickeners, lagoons, sedimentation basins and monofills. Conditions would likely be much different in a landfill due to the presence of organics, low pH, and redox conditions. Sludge was analyzed for total arsenic, iron, and aluminum initially, and after two and six months of storage. TCLP and Ca WET were conducted on dry (100 percent) sludge and a 20 percent solids concentration sludge. Lagoon supernatant, pH, dissolved oxygen, and redox potentials were measured at the same intervals, along with arsenic, iron, and aluminum.

Coagulation Residuals

Lagoon supernatant arsenic concentration was measured over time to demonstrate the effect of lagoon conditions on arsenic release from the solid to liquid phase (Figure ES.3). The data demonstrate that arsenic leaching from the lagoon solids occurred soon after lagoon simulation testing was initiated. All residuals, except for Lockwood, MT, demonstrated increased levels of arsenic in the supernatant. The total arsenic concentration in the supernatant very closely followed the trend of total iron release from the lagoon solids, and generally followed a lowering of the redox potential.

Lime Softening Residuals

The only lime softening residual used for lagoon simulation testing was from New Iberia, LA. The lagoon test results demonstrated that there was no release of arsenic to the lagoon supernatant under lagoon storage conditions. The lagoon pH remained at 11.7 and the redox potential after storage increased to as high as 178 mV. There were only small concentrations of iron and Al present in the lagoon supernatant residual stream initially, and no leaching over time was noted for either of these metals. Due to the chemistry of lime softening residuals, arsenic released under lagoon storage conditions is not expected to be a problem.

Modified Lagoon Conditions

The LADWP ferric residuals were used to develop two additional lagoon simulation tests which included a reduced pH lagoon condition and a lagoon with added biodegradable organic matter (BOM). These tests were intended to simulate more closely the conditions found in a sanitary landfill. Initial pH of the unaltered and BOM lagoons was approximately 7.0, while the reduced pH was less than five. The unaltered and BOM lagoon had reducing conditions at -100 to -200 mV, while the reduced pH remained redox positive. The lagoon supernatant arsenic data (Figure ES.4) demonstrate that the reduced pH and BOM additions to the LADWP residuals increased arsenic release above that of the unaltered lagoon.

Redox Potential

Figure ES.5 presents arsenic concentrations in lagoon supernatant collected at zero, two, four, and six months into lagoon simulation testing versus redox potential. For the majority of data points, redox potential became negative quickly, and was on the order of those found under reducing conditions in landfills. A key exception was the softening solids. Trends were similar for iron release (see Figure ES.6).

TCLP and Ca WET Results

Solids samples were collected from the lagoon simulations analyzed after two and six months of storage time for TCLP and Ca WET arsenic analyses. These data (using 100-percent solids samples) are shown in Table ES.4 along with the results from the fresh residuals sample analyzed prior to lagoon testing. Ca WET arsenic levels increased after two months of storage for all residuals except the lime softening residuals. TCLP arsenic levels were very low regardless of aging. Both tests were also conducted on sludge dried to a 20-percent solids concentration. Ca WET arsenic concentrations were generally higher for the 20-percent solids concentration samples than the 100-percent solids concentration samples.

Table ES.4

Lagoon TCLP and Ca WET results

	TCLP As (mg/L)			Ca WET As (mg/L)		
Utility name	Fresh	Aged 2 months	Aged 6 months	Fresh	Aged 2 months	Aged 6 months
Louisiana Water Company (New Iberia), LA	0.012	0.01	0.01	0.096	0.05	0.108
City of Great Falls, MT	0.158	0.048	0.109	4.83	6.11	7.18
City of Helena. MT	0.125	0.13	0.144	1.88	5.61	10.3
Dept. Of Public Utilities (Billings), MT	0.013	0.01	0.009	0.863	2.26	3.19
Lockwood Water Users Association (Lockwood), MT	0.011	0.009	0.013	0.224	0.26	0.266
Los Angeles Dept. of Water & Power (LADWP), CA	0.162	0.079	0.093	4.931	15.24	13.1
Indiana-American Water Company, IN	0.031	0.01	0.02	2.47	10.23	5.38

Note: TCLP and Ca WET test results in this table were conducted 100-percent solids concentration

The results demonstrate that each of the lagoon residuals was able to easily meet the TCLP arsenic limit. The maximum arsenic concentration noted was approximately 0.15 mg/L. The Ca WET arsenic data, however, present a different picture in terms of arsenic leaching. These data indicate that the fresh residuals would only be slightly below the 5.0-mg/L limit of the Ca WET test, and after lagoon storage for six months, four of the six residuals would exceed the 5.0-mg/L Ca WET arsenic limit.

Comparison of Extraction Tests to Lagoon Release

The Ca WET test caused a much higher arsenic release than the TCLP test for the lagoon simulations for almost all of the residuals tested. Both tests are designed to give some sense of leaching potential. TCLP and Ca WET arsenic concentrations in 20 percent residuals collected at two and six months are compared with measured lagoon release in Figures ES.7 and ES.8. At two months, CA WET arsenic levels were much higher (in some cases by more than two orders of magnitude) than both TCLP arsenic levels and measured arsenic concentrations in the lagoon supernatant. TCLP arsenic levels tracked much more closely than Ca WET concentrations with lagoon supernatant concentrations. At six months of lagoon storage, the results were similar.

The ratios of Ca WET and TCLP arsenic to measured lagoon arsenic are provided in Table ES.5. The Ca WET test arsenic concentration for the 20-percent solids concentration residuals was as much as 700 times greater than the lagoon arsenic release. The Ca WET test (100-percent solids concentration) measured arsenic release was three to seven times higher than the lagoon measured arsenic release. Overall, the Ca WET data measured using 100-percent solids demonstrated a much closer relationship to the lagoon release than did the 20-percent solids Ca WET results.

Table ES.5

Ratio of TCLP and Ca WET to arsenic lagoon release at two months

	Ca WET: Lagoon		TCLP: Lagoon	
Utility	20*	100*	20*	100*
New Iberia, LA	63	3.7	5.00	15.8
Great Falls, MT	20	3.3	2.30	0.05
Helena, MT	8	3.1	0.20	0.14
Billings, MT	156	6.6	1.90	0.06
Lockwood, MT	362	4.6	1.80	0.35
Los Angeles, CA	8	6.9	0.04	0.01
Indiana-American Water Company, IN	694	19.0	0.30	0.04

*Refers to percent solids concentration of the residuals used in the test

SFBW Clarification

A polymer screening analysis was used to select the two best polymers out of six polymers tested for each SFBW. In the screening tests, 100-ml residuals samples were dosed with 1-mg/L polymer and stirred for approximately 30 seconds to observe floc formation and turbidity was measured in the clarified water for each polymer screened. Results of SFBW settling tests using the best polymer conditions determined through screening are provided in Table ES.6. These data indicate that by conditioning with either $FeCl_3$ or polymer and gravity settling good arsenic removal was achieved. Successful removal of turbidity from SFBW suspended solids was also demonstrated to result in the removal of arsenic from solution.

Table ES.6

Comparison of best SFBW treatments using polymer and $FeCl_3$

	No chemical treatment	Polymer		Ferric chloride	
Utility name	Arsenic remaining (μg/L)	Dose (mg/L)	Arsenic remaining (μg/L)	Dose (mg/L)	Arsenic remaining (μg/L)
New Iberia, LA	30	1	7	120	5
Great Falls, MT	29	4	6	120	< 2
Helena, MT	159	4	< 2	20	3
Billings, MT	24	3	< 2	40	< 2
Lockwood, MT	10	3	4	40	< 2
LADWP, CA	50	0.5	< 2	20	7

TCLP arsenic concentrations were measured for each of the SFBW solids samples collected from the best chemical conditioning jar test. None of these tests resulted in TCLP arsenic concentrations near the regulatory limit of 5.0 mg/L.

Arsenic Residuals Stabilization

In order to determine if conditioning with lime would affect arsenic solubility, a series of tests was conducted using ferric chloride residuals from the LADWP. Additional tests were also conducted using caustic soda and soda ash additions to increase the ferric residuals pH and determine their relative effects on arsenic solubility.

Caustic Soda Conditioning Tests

Each residual was treated with caustic soda over the pH range encountered with lime treatment. This test was performed to determine whether the observed arsenic release was solely due to a pH effect. Total arsenic increased dramatically as the pH was increased with caustic soda (Figure ES.9). The same trends were observed for both the ferric and aluminum residuals. As much as 30 percent of the available arsenic was leached from the ferric residuals at pH 11.6 when no lime (calcium) was present. As the pH is raised, the hydrogen ion concentration decreases, shifting equilibrium and releasing arsenate to solution.

Soda Ash Conditioning Tests

Each residual was also treated with soda ash over the pH range encountered with lime treatment. This test was performed to determine the combined effect of pH and carbonate on arsenate release. The same trends were observed as with the caustic soda. Note that a maximum pH of 10.3 was obtained with soda ash. As much as 12 percent of the available arsenic was leached from the ferric residuals at pH 10.3 when soda ash was present.

Lime Conditioning Tests

The data from the lime conditioning tests yielded different arsenic releases than the caustic and soda ash conditioned residuals. For the ferric and alum residuals leachate, the arsenic was not appreciably released over the pH range evaluated (Table ES.7). The amount of lime added did not seem to affect the final concentrations. In all cases, each residual passed the TCLP test but failed the Ca WET test.

Table ES.7

Lime effects on ferric residuals

Lime percent (dry w/w)	pH	Arsenic leachate concentration (μg/L)	Arsenic leached (percent)
0.0	7.5	10	0.06
1.4	9.1	42	0.29
4.4	10.8	82	0.55
9.1	12.1	29	0.20
18.5	12.6	21	0.15

UTILITY DECISION TREE FOR ARSENIC RESIDUALS MANAGEMENT AND DISPOSAL

There are several treatment techniques that can effectively remove arsenic from drinking water. It is essential for utilities in the process of selecting an arsenic removal treatment technology to evaluate the types of residuals that would be generated, their expected arsenic concentrations, and pre-treatment strategies that would be required prior to final disposal. Possible pathways for residuals handling and disposal are shown in Figure ES.10.

The left side of the figure shows the arsenic removal processes that generate residuals streams with significant concentration of solids. For such processes, solids removal either with or without chemical addition is required as a first step to remove arsenic from the liquid phase. Secondary supernatant streams may or may not need additional treatment before they are acceptable for sewer or stream discharge, recycle, or reuse. Solids generally require thickening or dewatering and must pass the TCLP or Ca WET test to be disposed of in a non-hazardous waste landfill.

The right side of the tree illustrates the path utilities would likely follow upon choosing to remove arsenic from a liquid residual stream using disposable AA, Fe-based media adsorption or IX, which all generate liquid streams that do not contain appreciable levels of particulate solids. The arsenic in these streams can be removed using coagulation and clarification, or adsorption before being suitable for disposal.

Although not shown as a final disposal option, lagooning in impoundments or evaporation ponds (in the west) is a viable interim strategy, particularly for membrane concentrates and IX waste streams. Such storage or handling would of course be subject to regulation established by the statc.

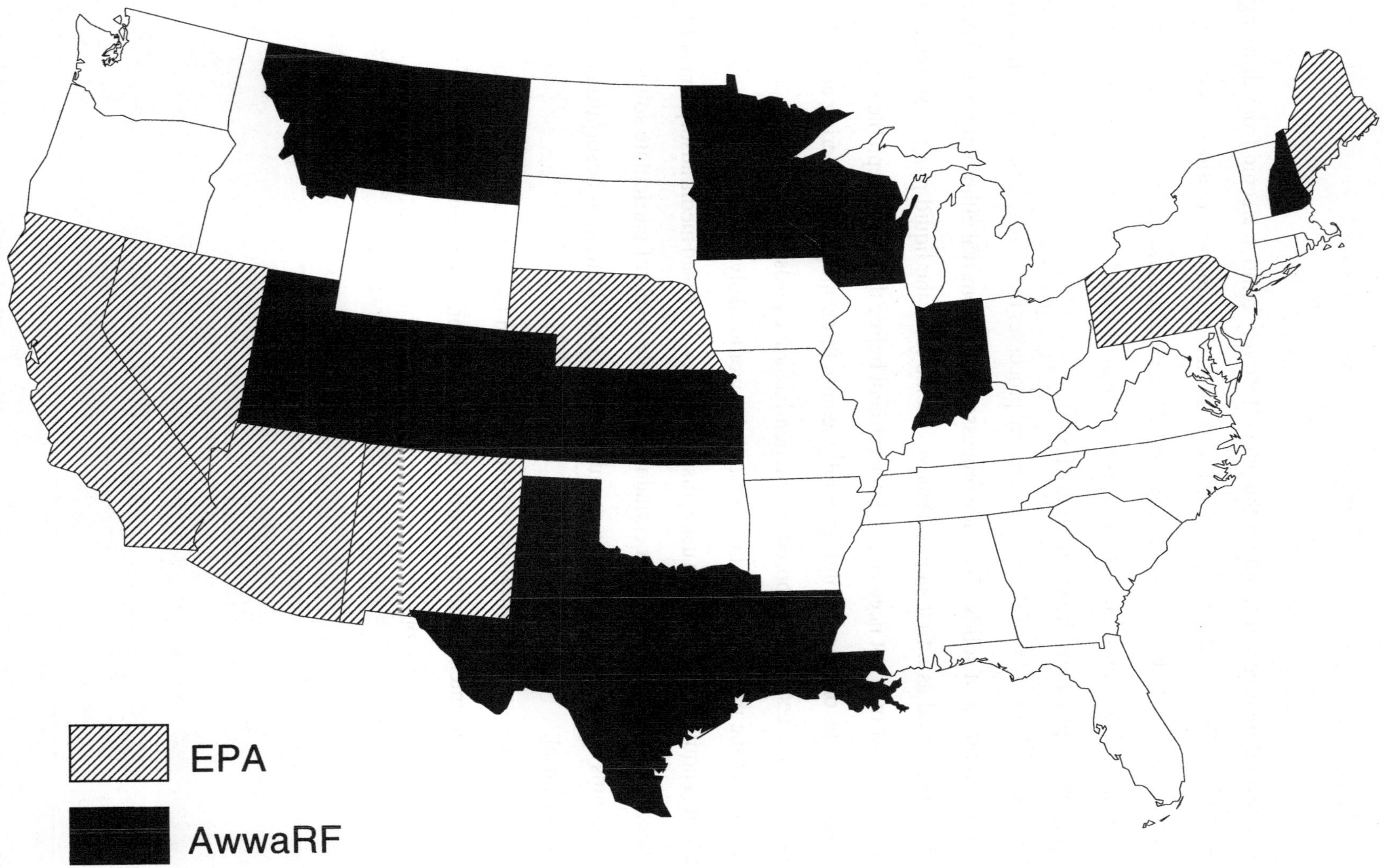

Figure ES.1 Regulatory survey states

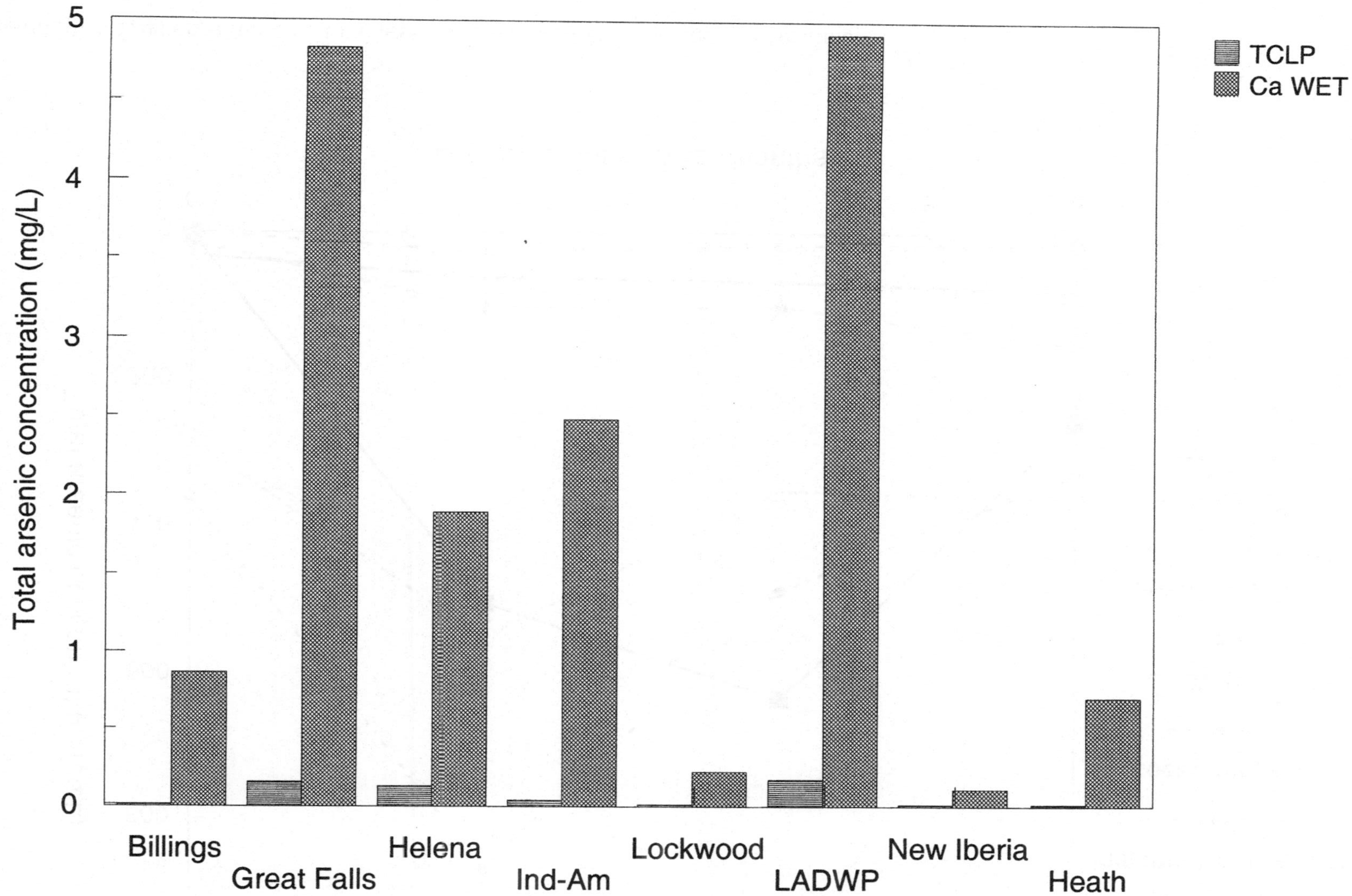

Figure ES.2 TCLP and Ca WET arsenic concentrations for WTP settled residuals solids

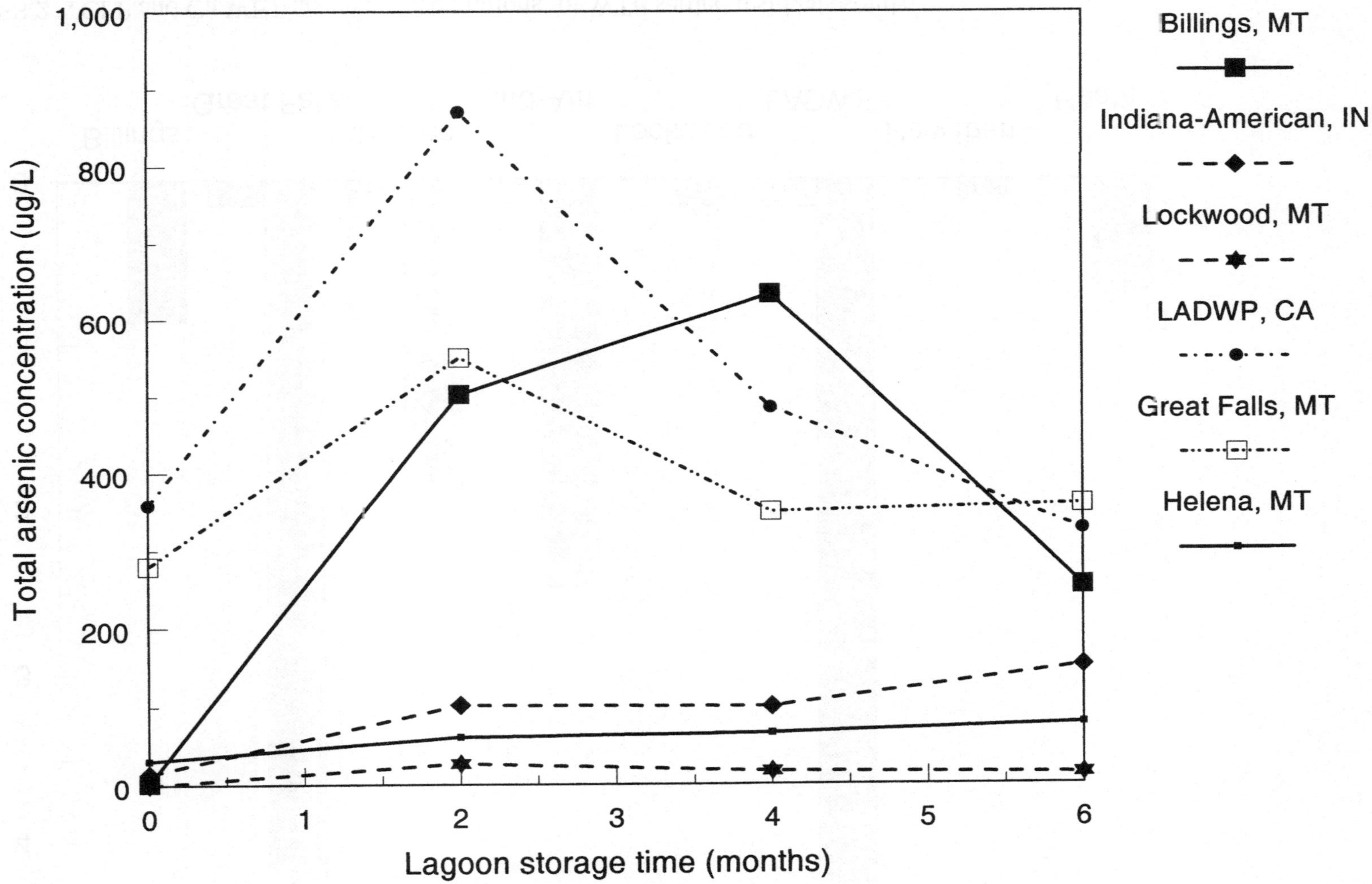

Figure ES.3 Lagoon simulation total arsenic concentration in the decant over time

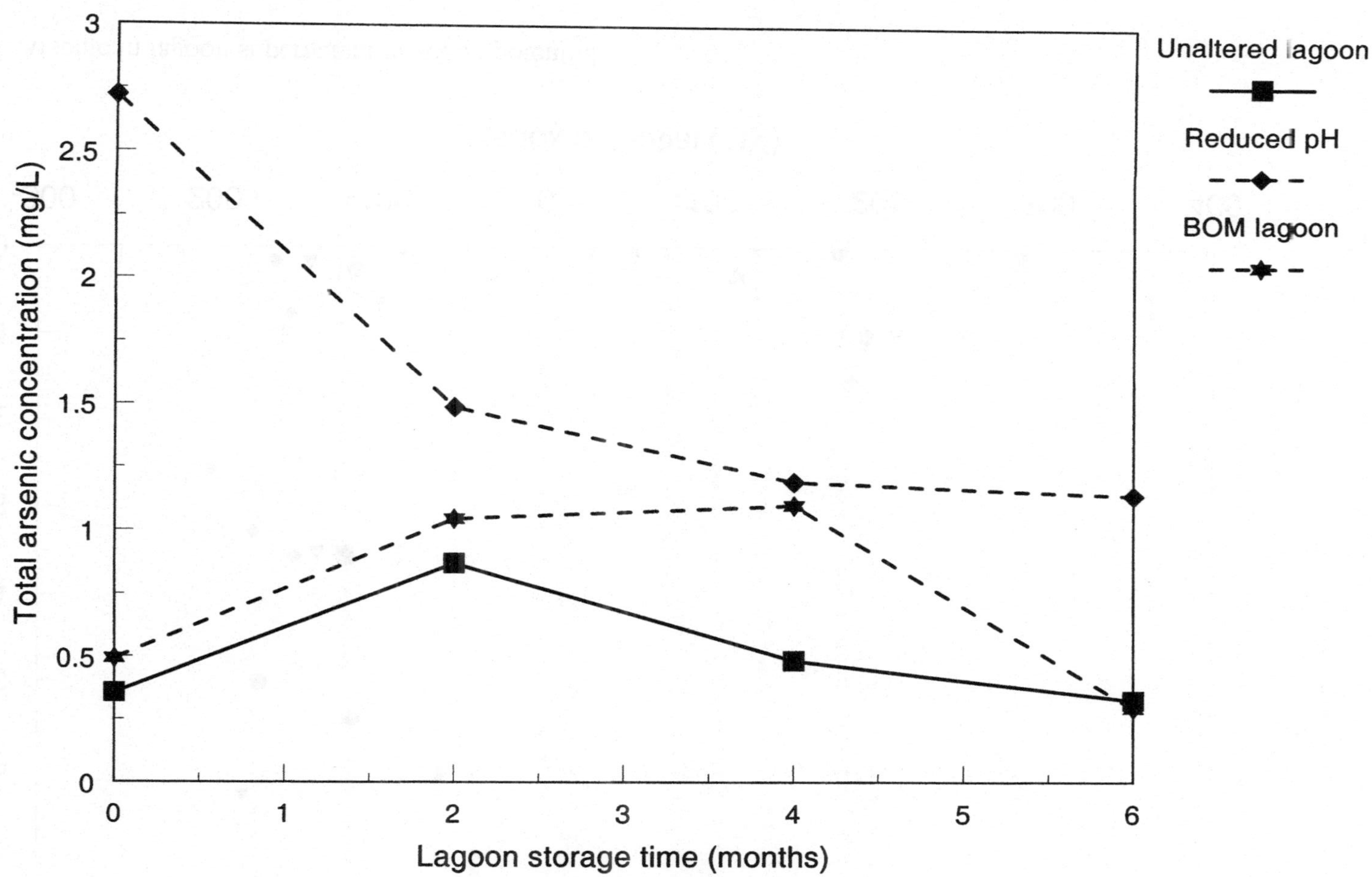

Figure ES.4 Total arsenic concentration in the lagoon simulation decant over time for LADWP ferric residuals in typical, BOM, and reduced pH lagoons

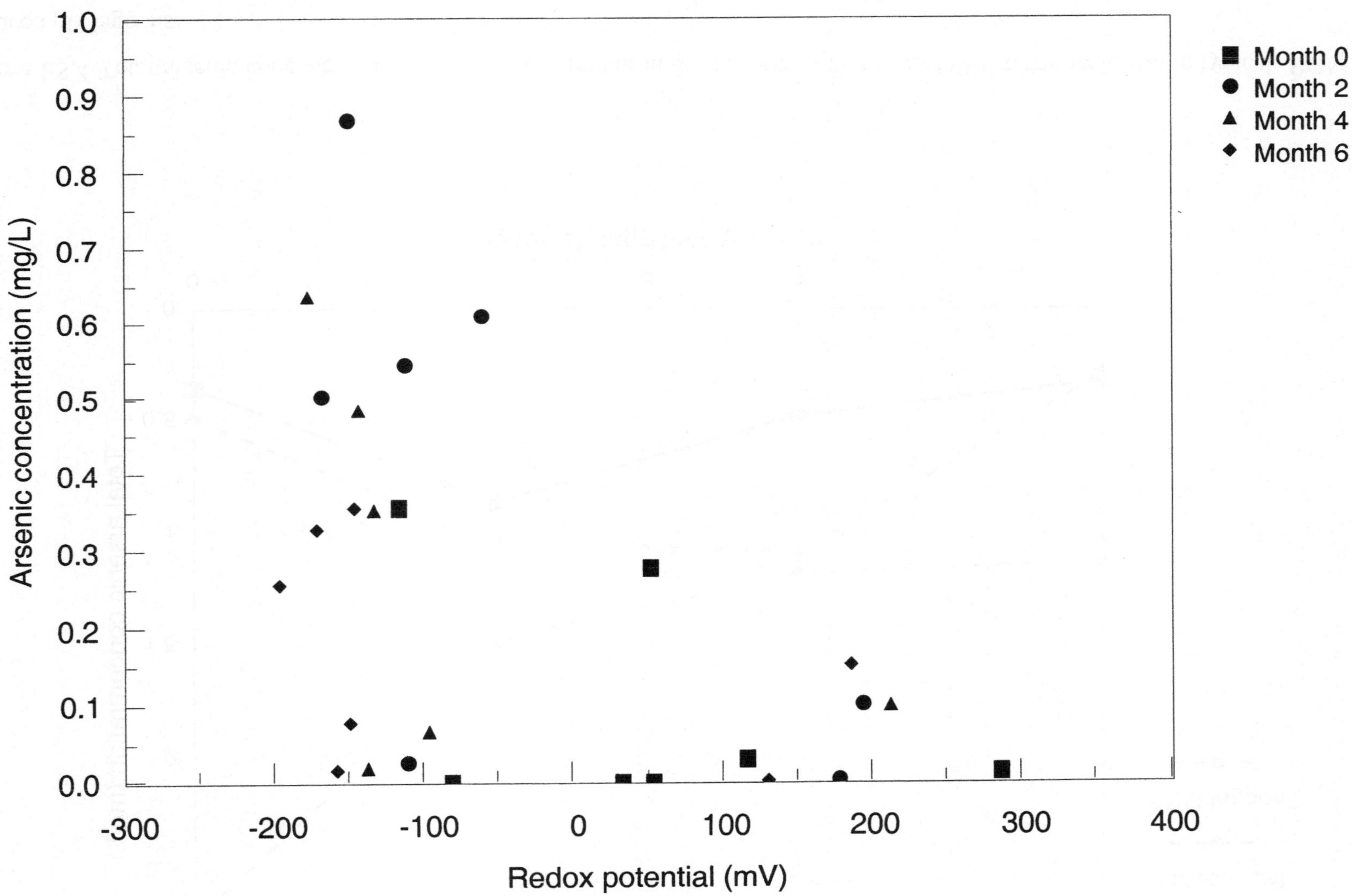

Figure ES.5 Arsenic in lagoon supernatant vs redox potential

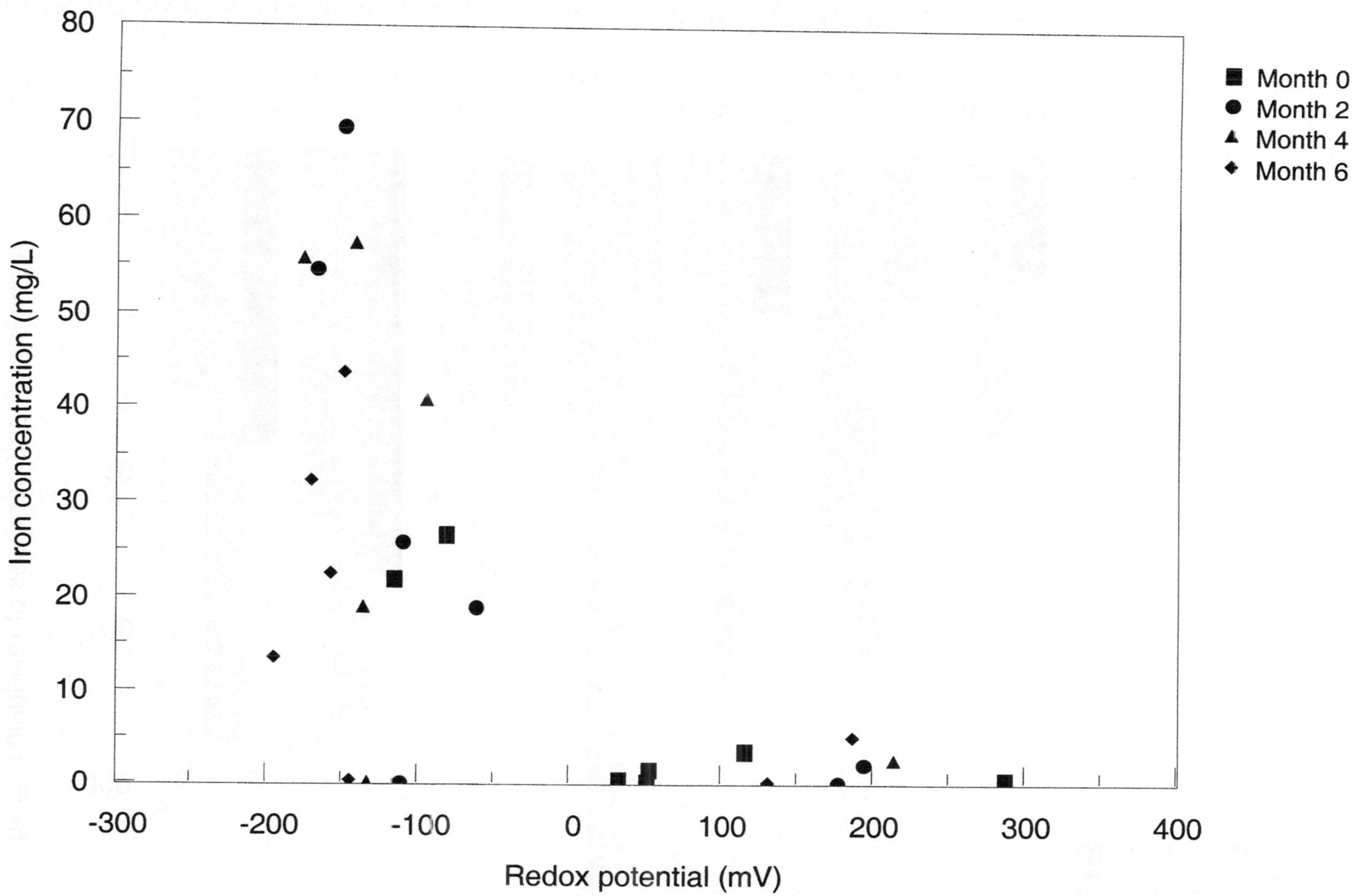

Figure ES.6 Iron in lagoon supernatant vs redox potential

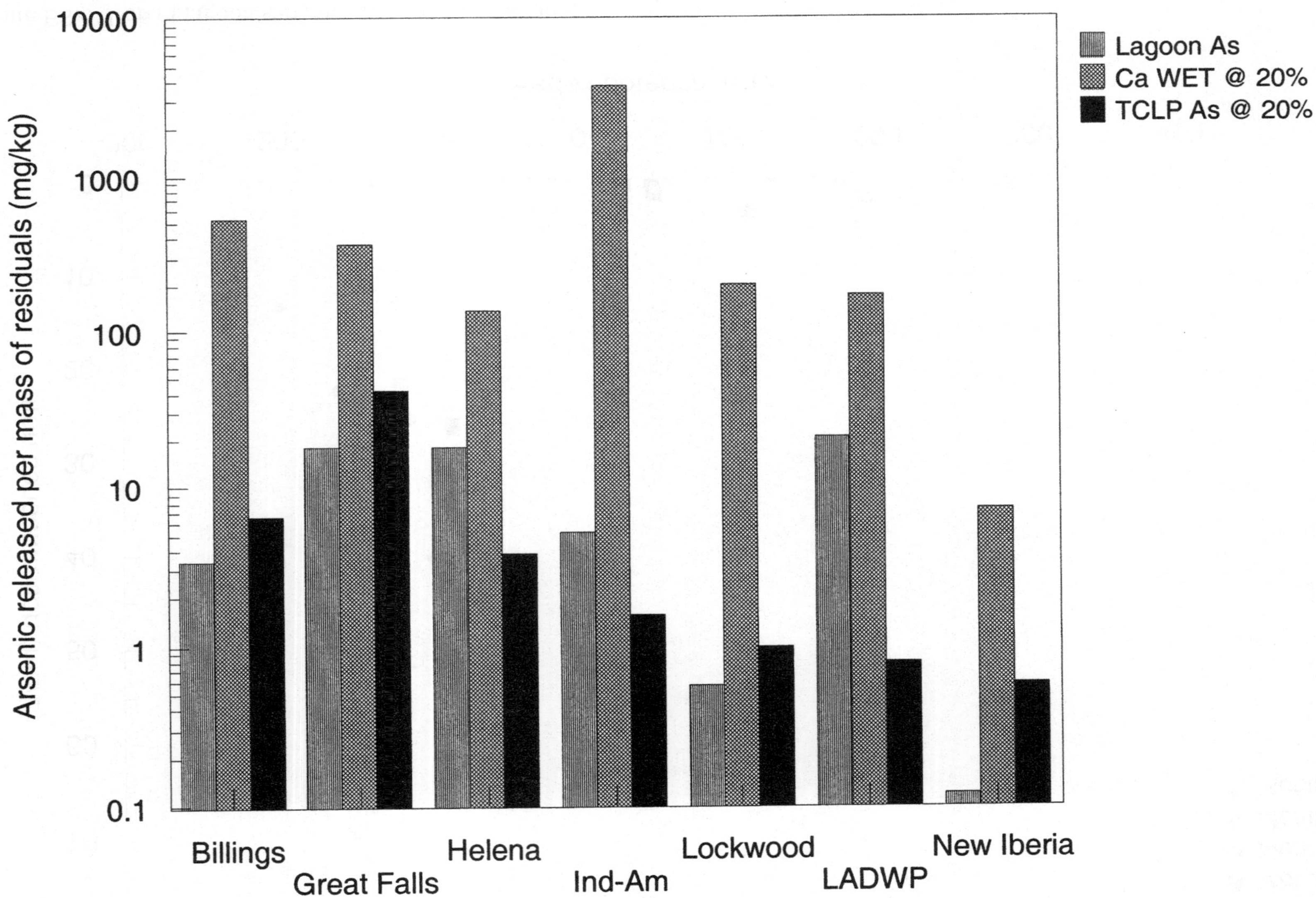

Figure ES.7 Comparison of measured lagoon arsenic release with Ca WET and TCLP at two months

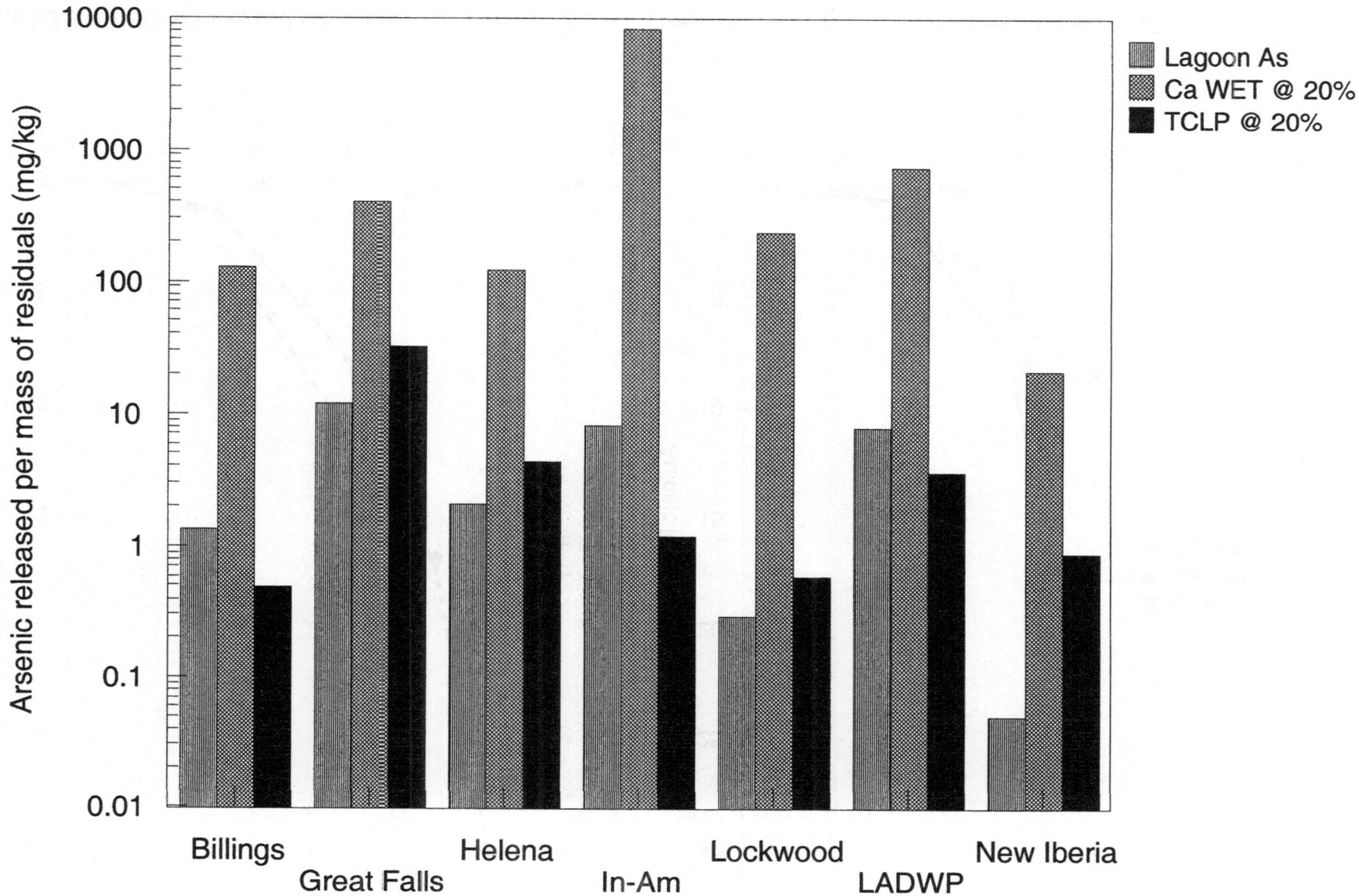

Figure ES.8 Comparison of measured lagoon arsenic release with Ca WET and TCLP at six months

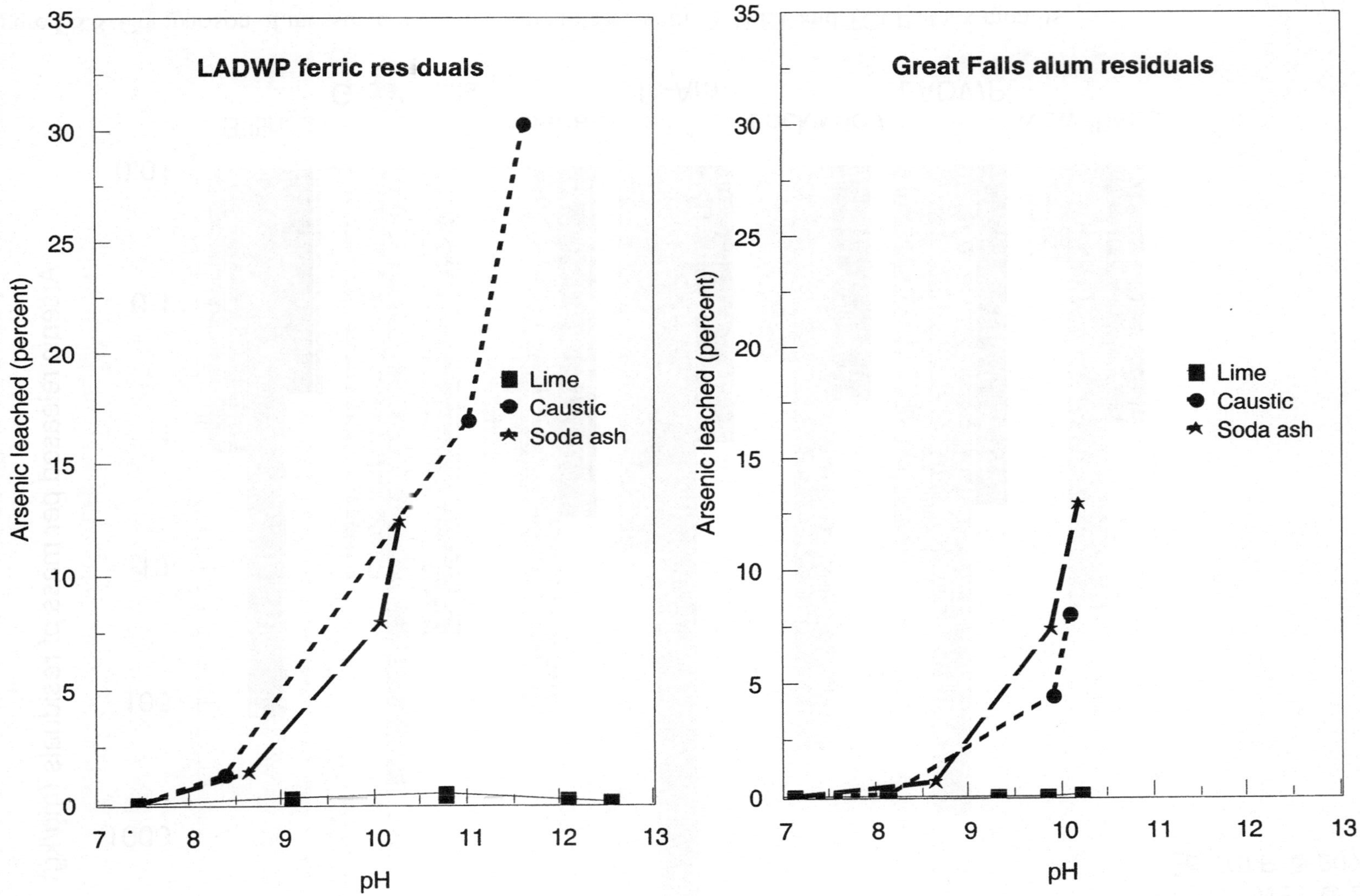

Figure ES.9 Effect of lime, sodium hycroxide, and soda ash on arsenic leachability from ferric and alum residuals

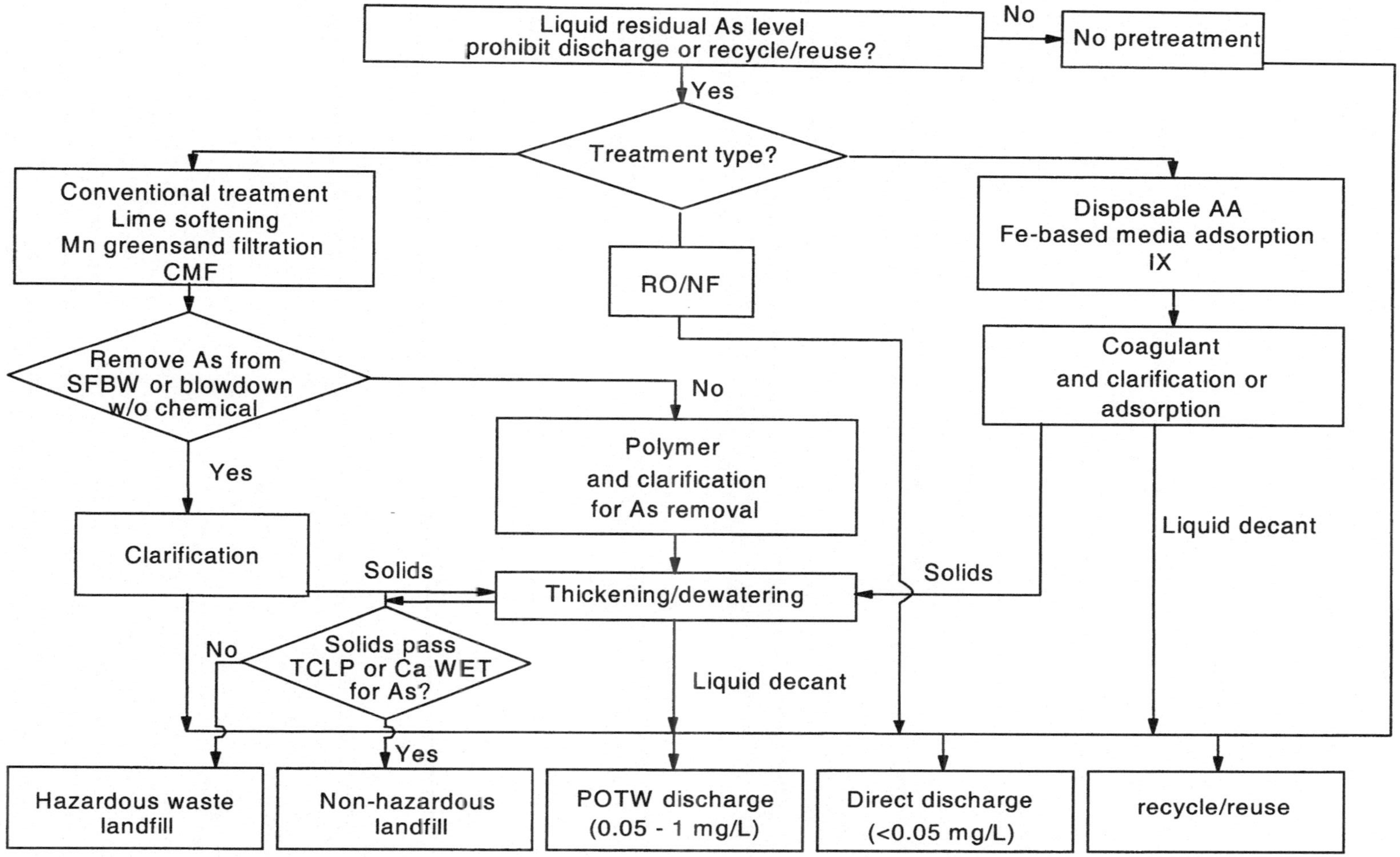

Note: Spent media disposed of in non-hazardous waste landfill

*AA and iron-based media adsorption backwash water expected to meet POTW direct discharge or recycle arsenic criteria

Figure ES.10 Arsenic residuals handling and disposal decision tree

CHAPTER 1
PROJECT INTRODUCTION

BACKGROUND NEED

Arsenic (As) has been recognized as a human health concern for many years. Because of this concern, a number of studies were conducted, primarily in the 1970s, to evaluate arsenic removal by conventional (coagulation, softening) water treatment processes. These studies found that both coagulation and softening processes could remove arsenic to levels considered acceptable, at that time, for protection of human health.

Arsenic in residuals can come from two major sources, the raw water and the treatment chemicals. Based on recent surveys by Frey and Edwards (1997), locations in the U.S. that are likely to have high raw water arsenic levels are known. Cornwell and Koppers (1990) published data about arsenic contamination of iron and alum coagulants, and the subsequent impacts on residuals disposal alternatives. Therefore, as utilities remove more arsenic from the source water, they must consider residuals management issues and costs.

The handling and disposal of arsenic-laden residuals may be a problem because various handling and disposal methods can release arsenic back to the environment. Because arsenic removal is sensitive to both the pH and the oxidation state, any process that changes pH or results in a reducing environment may release arsenic from the solid phase. These processes include chemical conditioning during dewatering, storage and lagooning, and ultimate disposal options such as landfilling, land application, indirect discharge to a sanitary sewer, and coagulant recycle.

With regard to arsenic removal from drinking water, a number of papers have been published detailing the removal effectiveness of both conventional treatment (coagulation and softening) and special processes (ion exchange, reverse osmosis and activated alumina). The technologies for arsenic removal have been summarized by Amy and Edwards (1999). Both aluminum (Al)and iron (Fe) salts are effective for arsenic removal, but both the pH and oxidation state of the arsenic are important factors in determining the removal efficiency. Softening is also effective and the removal mechanism appears to be by $Mg(OH)_2$ (Amy and Edwards 1999). Therefore, the pH of precipitation and the raw water chemistry are important considerations for arsenic removal by softening.

There are areas where additional studies will be important with regard to the impact of arsenic removal from residuals. If modifications of the coagulation processes are needed to meet the lower required levels of arsenic in finished waters, this may impact residuals handling in ways not yet understood or anticipated. There are many questions regarding differences in raw water quality and the retention of arsenic on residuals:

- How much does arsenic in coagulants add to the problem of arsenic in residuals for those facilities that must also remove source water arsenic?
- Will arsenic associated with coagulant solids be reduced from the V to the III oxidation state during storage and be released from solids?
- Will arsenic be released from residuals during pH changes that might occur during conditioning, storage, or lagooning?
- Are there techniques that can be used to reduce the mobility of arsenic during handling and disposal?
- What factors (such as arsenic content, coagulant type, residuals characteristics) influence the release of arsenic during toxicity testing?
- Do solids produced from arsenic removal processes fail TCLP or Ca WET tests?

A review of the literature suggests that none of these questions can currently be answered. One would suspect that release of arsenic from $Mg(OH)_2$ solids would occur if the pH was lowered and the magnesium solubilized. The TCLP data for softening residuals reported by Amy and Edwards (1999) do not indicate significant amounts of arsenic release. However, this is primarily because the softening residuals have a greater buffering capacity than coagulant residuals. Data from Amy and Edwards (1999) showed that two different ferric hydroxide coagulant residuals with similar arsenic levels released very different amounts of arsenic in the TCLP test. Cornwell et al. (1992) conducted research on leaching of arsenic from coagulant residuals subject to acid rainfall and also found varying releases for different residuals. Also, the impacts of variations in residuals characteristics caused by raw water quality differences, aging, and differences in coagulation practices on the release of arsenic dewatering or the TCLP extraction test have not been studied. Because coagulation can occur by several mechanisms (sweep and charge neutralization) and coagulants can differ in form (polyaluminum chloride vs alum), differences in the retention of arsenic would not be unexpected.

The oxidation state of arsenic is also critical in determining release. Reduction from the As(V) to the As(III) oxidation state under anaerobic conditions that could occur during storage, might be important for predicting arsenic release. However, no data are available indicating that these changes in oxidation state of arsenic occur and that they cause the release of arsenic. One would expect to find reduced environments, especially in lagooned residuals but the impact on arsenic solubilization has not been investigated. Similarly, the carbon dioxide by-product (CO_2) generated by biological activity in lagooned softening residuals could solubilize magnesium hydroxide ($Mg(OH)_2$), and as a result release arsenic.

Finally, the presence of arsenic in treatment chemicals must be considered. A survey of ferric chloride coagulants from eight plants found a range of 108 to 122 mg As/kg liquid coagulant. Research in the Netherlands (Cornwell and Koppers 1990) has shown that up to 40 percent of the arsenic in their residuals can come from the iron coagulants. Other research projects (Cornwell et al. 2002) measured contaminants in drinking water coagulants and determined that the impacts on drinking water and residuals are not significant. Utilities that do experience an arsenic problem in their residuals should evaluate both the raw water and chemical additions to determine arsenic levels. If chemical additives are found to contain high concentrations of arsenic, then utilities may need to find alternative chemicals sources.

OBJECTIVES

The objective of this research project was to better understand the factors that may cause the release of arsenic from solid residuals into the environment. The project tasks were structured in order to determine arsenic release from residuals due to chemical conditioning for dewatering, thickening, short- and long-term storage, or disposal applications such as landfilling, discharge to a wastewater treatment plant, and land application. The project findings were used to investigate innovative methods for preventing arsenic mobility from residuals into the environment. These research efforts will help provide the data necessary to develop recommendations for regulators and utility managers that will help to minimize arsenic release during handling, storage, and disposal of WTP residuals.

CHAPTER 2

REGULATORY ISSUES

BACKGROUND

The removal of arsenic from drinking water is of interest because of the potential impact of arsenic on human health. Ambient water quality limits have been set for arsenic based on fish consumption (0.14 μg/L) and both drinking and fish consumption (0.0175 μg/L) for the same water. The arsenic water quality criteria are for a 10^{-6} cancer risk. The SDWA MCL for arsenic was 50 μg/L but has been lowered to 10 μg/L.

The ambient water quality criteria from Amy and Edwards (1999) are listed below

Water classification	Total arsenic	Trivalent arsenic
Fresh water acute (μg/L), dissolved		360
Fresh water chronic (μg/L), dissolved		190
Marine water acute (μg/L), dissolved		69
Marine water chronic (μg/L), dissolved		36
Human health - fish consumption (μg/L)	0.14	
Human health - water and fish consumption (μg/L)	0.018	
SDWA drinking water criteria (μg/L), total	10	

One consequence of the human health fish consumption limit set under the Clean Water Act (CWA) is that effluent discharges from lagoons and filtrates from dewatering processes may exceed effluent limits. The arsenic limits will likely be set based on the ability of the stream to assimilate the arsenic and remain below the arsenic standard for the stream. Discharge limits may also exist for indirect discharge of arsenic-laden residuals, overflows, and filtrate. These limits are based on Technically based local limits (TBLLs) of the Industrial Pretreatment Program. The residuals must meet established TBLLs if they are to be discharged to the sanitary sewer. Most Industrial

Protection Programs have TBLLs in the range of 50 μg/L to 1,000 μg/L. The concentration of arsenic in sedimentation sludge blowdown is highly site specific depending on the concentration of arsenic removed, the amount of coagulant used, and the removal efficiency of the treatment process.

Land application of sewage sludge must conform to 40 CFR part 503 regulations. To meet the EPA "clean sludge" limits, arsenic must be below 41 mg/kg. If this limit is exceeded application must be limited to a cumulative total of 41 kg/hectare. Calculations made by Amy and Edwards (1999) indicate that this could impact indirect discharge of arsenic-laden WTP residuals into the sewer depending on the background arsenic in the wastewater.

For landfills, the applicable criteria for non-hazardous disposal is the toxicity characteristic leaching procedure (TCLP). If arsenic in residuals leachate exceeds the TCLP limit of 5 mg/L, the residuals are characterized as hazardous. The current TCLP limit is based on a leachate concentration that is 100 times the former drinking water MCL of 50 μg/L. In California, the California Wet Extraction Test (WET) is used in place of the TCLP test. Each of these toxicity determination tests is described briefly in Chapter 3.

Data collected by Amy and Edwards (1999) for extractability of arsenic from residuals shown in Table 2.1 indicate that all residuals would meet the current TCLP limit. For the California WET test, failure would be more likely since the WET test extraction from Utility O was more than 10 times higher than results from the TCLP test.

In summary, with regard to regulations, the residuals issues due to a lowered arsenic MCL requires further attention based on these data. Filtrates and overflows may limit disposal to streams, residuals discharge to sewer systems may exceed TBLLs because of their impact on land application of biosolids, and some residuals may be considered toxic (fail the TCLP test).

Table 2.1

TCLP and Ca WET results for conventional WTP arsenic residuals

		Total digested concentration*		TCLP extract†		
Sample	Type of residual	mg Fe/kg dry solids	mg As/kg dry solids	Percent dry solids‡	Fe (mg/L)	As (μg/L)
Utility F Coagulation residuals	Lime softening	18,624	2.4	54.5	0.1	0.9
Utility F Softening residuals	Lime softening	9,945	6.9	54.7	0.2	3.9
Utility F SFBW	Lime softening	162,000	35.2	0.1	0	1.4
Utility G	Lime softening	7,536	14.8	41.8	0	2
Utility J	Lime softening	3,667	24.6	87	0.1	28.4
Utility L	Alum coagulation	N/A	N/A	0.34	N/A	9.3
Utility C	Fe/Mn removal	37,345	369.0	66.3	2.4	44.4
Utility O	Iron coagulation	19,350	337.8	100	182.4	1,559.6
Utility O, WET extraction§	Iron coagulation	19,350	337.8	100	1,397	18,500

Source: Amy and Edwards 1999
*Diluted 1:1000, digested with 5 percent (v/v) HNO_3, sonicated at 45°C for 30 min
†Extracted with glacial acetic acid according to the EPA standard method
‡If dry solids <0.5 percent, then TCLP extract = liquid portion of the sample
§State of California test, extracted with citric acid
N/A indicates not analyzed due to insufficient amount of solid sample

PURPOSE

The purpose of this chapter was to summarize the existing state and federal regulations that apply to the handling, transportation and disposal of As-laden residuals. It is intended to be a continuation of the state regulatory summary generated by the EPA in the report titled "Regulations on the Disposal of Arsenic Residuals" (USEPA 2000). The EPA report summarized the regulations from seven U.S. states: CA, AZ, NM, NV, ME, PA, and NE. This report provides state regulations from 10 additional U.S. states: MN, KS, LA, CO, IN, WI, UT, MT, NH and TX. The EPA and AwwaRF states evaluated are shown in Figure 2.1.

FEDERAL REGULATORY SURVEY

General

Although there are no federal regulations that specifically deal with the disposal of water treatment plant residuals, a number of federal regulations are in place that affect specific methods of disposal of these materials. In general, these regulations are meant to protect either the receiving water or land and underlying groundwater at the location of the disposal site. This section provides a general overview of these regulations.

Residuals from water treatment plants are disposed of according to the physical form of the residual; that is, if the residual is a liquid or a solid. Liquid residuals can be disposed of by means of direct discharge, indirect discharge, or underground injection. In some cases, liquid residuals can also be disposed of in a landfill or monofill. Solids are typically disposed of in monofills, industrial or mixed use landfills, municipal solid waste landfills, hazardous waste landfills, through land application, or by reuse. Lagoons can also be used to manage these residuals. Each of these disposal methods is regulated by different requirements and regulations as described in this section.

Most states have the administrative authority (primacy) to implement these federal regulations, and so are required to establish and administer regulations that meet the requirements of these acts. The regulation of waste disposal is primarily the responsibility of the states, but the state regulations must meet or exceed the federal regulations.

Liquid Residuals

Direct Discharge to Surface Waters

The Clean Water Act (CWA), 33 USC s/s 1251 et seq. (1977), a 1977 amendment to the 1972 Federal Water Pollution Control Act, regulates the discharge of pollutants to waters of the United States. The Act was amended in 1978, 1980,1981 and 1987, and is meant to ensure that water bodies maintain the appropriate quality for their intended uses, including navigation, swimming, fishing, agriculture, and drinking water supply.

The CWA makes it unlawful for any person to discharge any pollutant into navigable waters or wetlands unless a National Pollutant Discharge Elimination System (NPDES) permit is obtained under the Act. The permit specifies the permissible concentration or level of contaminant in a facility's effluent. Although this is a federal regulation, its implementation and enforcement is usually carried out by the states, under EPA authorization. This federal regulation requires acceptance or assumption of primacy by the individual states, otherwise, the program is administered by the regional EPA office. In states where authorization or primacy has not been given by the EPA, the EPA regional offices grant the NPDES permits. The CWA provides for the delegation by EPA of many permitting, administrative, and enforcement activities of the law to state governments. In states with the authority to implement CWA programs, EPA still retains oversight responsibility.

State regulations and guidelines controlling the discharge of residuals, however, can vary throughout the United States. In all states allowing direct discharge, there are limitations on suspended solids concentrations that can be released. Some states do not allow direct discharge of residuals even if residuals pretreatment is performed to remove contaminants.

The CWA gave the EPA the authority to set effluent standards on an industry basis (technically-based) and continued the requirements to set water quality standards for all contaminants in surface waters. Under the CWA, drinking water plants that directly discharge residuals must have an NPDES permit and may only discharge pollutants in conformance with the terms of that permit. Generally, each permit must include technically-based water quality effluent limits if such limits have been developed for each different industry. To date, EPA has not developed categorical standards for water treatment plants. Therefore, such permit limits are generally based on best professional judgement (BPJ). Under the CWA, arsenic is specifically identified as a toxic pollutant, and so a BPJ-based effluent limit must be included as a condition for an NPDES permit for any residuals containing arsenic.

Federal guidelines for controlling drinking water plant effluents were drafted by the EPA in 1994, but never promulgated. The draft guidance specified BPJ control technologies for different types of plants depending on treatment used. The discharge limits, however, only specified limits on pH and total suspended solids (TSS). No limits on the toxic pollutants such as arsenic were suggested.

EPA has established numeric water quality criteria under the authority of the CWA, for a number of parameters, including arsenic (40CFR131.36). For acute levels of arsenic in fresh waters, a recommended limit of 0.36 mg/L was set, and for chronic limits, a level of 0.19 mg/L was set. For salt waters, the acute level was set at 0.069 mg/L, and the chronic level was set at 0.036 mg/L. These criteria are used by the regulatory agency (either EPA or state) to determine their water quality standards. Generally, states must develop numeric surface water quality standards for the priority pollutants where a discharge or the presence of the pollutant could reasonably be expected to interfere with the designated uses of a receiving water body.

Indirect Discharges to a Sanitary Sewer

In some cases, a drinking water treatment plant may be able to dispose of its residuals to a sanitary sewer. In this case, an NPDES permit would not be required of the drinking water plant. However, if the sewer system is part of a publicly owned treatment works (POTW), the drinking water facility would need to meet the TBLL of the POTW. This requirement comes from the provisions of the CWA pretreatment requirements. EPA requires that all wastewater plants comply with pretreatment requirements, and this rule clearly specifies that in this case, water plant residuals should be classified as industrial wastes. Pre-treatment requirements for the residual may be specified, including limits on the amount of arsenic, since the arsenic could impact the operation of the POTW. In general, a discharge to a POTW is only acceptable if the arsenic concentration falls within the established Technically Based Local Limits (TBLL) of the Industrial Pretreatment Program. TBLLs are individually determined for each POTW, and take into account background levels of contamination in the municipal wastewater. In some cases, the arsenic limit may be based on the limits imposed on the POTW in their land application of wastewater biosolids under 40 CFR 503.

For wastewater biosolids, 40 CFR 503 specifies allowable limits for the arsenic concentrations as a function of disposal methods. POTWs utilizing land application are subject to the Land Disposal Limit, Land Application Ceiling Limit, and Land Application Clean Sludge Limits which are 73 mg/kg, 75 mg/kg, and 41 mg/kg, respectively. If the arsenic concentration in the biosolids exceeds the clean sludge criterion, land application is limited to 41 kg per hectare (36.6 lb/acre). As a result, most TBLLs are based on the clean sludge criteria. For example, the typical POTW removal efficiency for arsenic is approximately 45 percent. Assuming biosolids production is around 1,200 pounds per million gallons of water treated, the maximum allowable headworks loading will be around 0.109 pounds of arsenic per million gallons of wastewater treated. This equates to a total influent arsenic concentration of around 13 μg/L. As a result, if a water system has a background arsenic concentration near 13 μg/L, it may not be possible to discharge to the sanitary sewer if the TBLL is based on the POTW's biosolids disposal permit (Amy and Edwards 1999).

Underground Injection

Liquid residuals may be disposed of by means of underground injection. This disposal method would be regulated through the Underground Injection Control (UIC) program of the Safe Drinking Water Act (SDWA). Federal UIC regulations prohibit the subsurface discharge of fluid through a well or hole whose depth is greater than its width without a permit. Underground disposal of liquid residuals from a drinking water facility would be through Class V wells unless the material was designated as hazardous. For hazardous material injection, the designation of Class I well is used. These two types of wells are characterized as:

- Class I includes wells used to inject hazardous waste or industrial and municipal waste beneath the lower-most formation containing an underground source of drinking water within 1/4 mile of the well bore

- Class V includes the injection wells that are not considered in Class I, are not oil and gas wells, and are not involved in mining. Class V injection wells are typically shallow disposal systems for placing nonhazardous fluids underground, and have been regulated under the Safe Drinking Water Act since 1983 to protect underground sources of drinking water

In general, underground injection is prohibited where it would cause any underground source of drinking water to exceed any SDWA mandated drinking water standard, or otherwise impact public health.

Solids Residuals

Landfilling of Nonhazardous Material (RCRA Subtitle D)

Under the Resource Conservation and Recovery Act (RCRA) Subtitle D regulations (40 CFR Parts 257 and 258), criteria have been established for the design and operation of nonhazardous, solid waste landfills. Landfills that receive only drinking water treatment residuals (monofills) are subject to the requirements of 40 CFR Part 257. These same criteria are used to evaluate any landfill that accepts non-hazardous, non-household solid waste, and are strictly performance based. Landfills that accept municipal solid waste are subject to the criteria of Part 258, which in addition to the performance based criteria of Part 257, contains design and operation criteria. RCRA contains requirements that restrict the location, operation and design of these landfills, and also sets requirements for groundwater monitoring and closure requirements. These two types of landfills—monofills and commercial nonhazardous landfills—are the most commonly used methods of disposal of water plant residuals in the country.

Landfill requirements under both Parts 257 and 258 specify that residuals be in a solid form with no free flowing liquids. Sanitary landfills also require that a material pass the Toxicity Characteristic Leaching Procedure (TCLP). The TCLP test, which includes analysis for a number of compounds including arsenic, is used to determine whether the waste should be characterized as toxic. Toxicity is one of four characteristics examined to determine if the waste is hazardous. This toxicity test is the primary area of concern for As-containing wastes. The current TCLP limit for arsenic is based on the former drinking water MCL, and is set at 100 times that level, or 5 mg/L. While the SDWA mandated MCL for arsenic has recently been lowered to 10 µg/L, the TCLP levels currently remain at 5 mg/L.

Subtitle D also requires management practices at these landfills to protect the environment. One of these mandates is that the landfill leachate not contaminate an underground water source beyond the solid waste facility boundary. Any release of these compounds to the groundwater that exceeds the permitted level would be considered contamination.

It should also be noted that EPA now prohibits the disposal of non-containerized or bulk liquid wastes in municipal solid waste landfills. RCRA requires that the landfill operator or owner determine whether the wastes are "liquid" according to the Paint Filter Liquids Method 9095. In this test, the wastes are placed on a #60 mesh filter paper. If any liquid passes through the filter within five minutes, the waste is considered a liquid. This requirement would impact the disposal of conventional treatment plant spent filter backwash water, for example.

Landfilling of Hazardous Material (RCRA Subtitle C)

Under RCRA Subtitle C, the landfilling of materials that are characterized as hazardous is regulated. As previously discussed, four criteria are used to determine if a material is "hazardous," including toxicity, ignitability, corrosivity, and reactivity. Water treatment residuals that exceed the TCLP limit for arsenic fail the TCLP test, and so would be deemed hazardous material because of toxicity.

Hazardous waste landfills are generally regulated by the federal government under the authority of RCRA, but some states also have this authority. Hazardous waste landfills are required to be permitted under 40 CFR 270, which specifies landfill construction and operation criteria, and are designed to isolate hazardous contaminants from the environment.

Bulk liquid waste disposal is not commonly permitted in landfills. If any water treatment residual contains free liquids as measured by the Paint Filter Liquids Test, (SW-845, Method 9095) it cannot be landfilled without being stabilized or treated by another method to remove free liquids.

In addition to these requirements, RCRA requires extensive documentation and manifestation of the material. Transporting of the material once it is deemed to be hazardous is also regulated.

On-Site Lagoons

Lagoons are not considered an ultimate disposal method for drinking water treatment residuals, because they are not a permanent disposal and will eventually require cleaning and final disposal of the solids. However, many utilities choose to manage some liquid residuals by storing them in lagoons on site, allowing for decantation and some evaporation. If the material stored in these lagoons is non-hazardous, then the management of the lagoons would be regulated by the Safe Drinking Water Act (SDWA.) The requirements of that act specify that states must establish programs to protect wellhead areas; that is, areas around a well or a public water supply wellfield from contaminants that may pose adverse effects on human health.

If the water treatment residual has failed the TCLP test because of arsenic concentrations, then on-site lagoons would be regulated by RCRA. Under RCRA, EPA has established criteria that prohibit practices that contaminate surface or ground water (40 CFR 257) and also established comprehensive design and operation standards applicable to surface impoundments.

Beneficial Reuse

Many drinking water treatment residuals are disposed of through land application. This practice is governed by RCRA, either under Subtitle C (hazardous material) or Subtitle D (non-hazardous material). In this set of regulations, land application would be considered "reuse."

If a waste failed the TCLP test for arsenic and was characterized as hazardous, the residual could be land applied under a RCRA 40 CFR 266 exemption. Under this exemption, the material could be land applied, but only after it had undergone a chemical reaction such that it was rendered non-hazardous.

For nonhazardous materials, there are very few federal regulations governing potential land application. The RCRA Criteria of Subtitle D section 257 establishes some general provisions governing the management of these operations. In general, the requirements address location in a floodplain, protection of endangered species, protection of surface water, land application to food chain crops, minimization of disease vectors, protection of air quality, and limits on explosive gases.

It should be emphasized that drinking water residuals are specifically exempted from the requirements set forth in 40 CFR 503(b) of the CWA. That regulation sets limits on the concentrations of toxic materials, including arsenic, allowed in land applied biosolids, but these limits do not apply to water treatment residuals.

STATE REGULATORY REVIEW

General

The state regulatory survey conducted during this research project included a total of ten states as follows:

- Utah
- Colorado
- Montana
- Kansas
- Minnesota
- Wisconsin
- Indiana
- Texas
- New Hampshire
- Louisiana

The regulatory survey focused on disposal of both liquid and solid phase residuals. Disposal of liquid residuals could include direct discharge to surface water by NPDES permit, sewer discharge, underground injection (UIC Program), or land disposal. Solid residuals disposal outlets could include a solid waste landfill, lagoon storage or stockpiling, or possibly a type of beneficial re-use such as land application, composting, or other.

The majority of the information used for development of these regulatory summaries was acquired by searching each of the state agency's internet home pages, by phone interviews with state regulatory personnel, and through professional contacts within the water industry. Many of the states investigated do not have pre-set regulatory limits for disposal of arsenic-laden residuals and thus regulate on case-by case basis.

Utah

Liquid Residuals

Direct discharge to surface waters. Utah is authorized by the EPA to administer the entire NPDES program through the Utah Water Quality Board. The Utah Pollutant Discharge Elimination System (UPDES) has as its authority R317-8 of the Utah Administrative Code, last revised in August of 1995. The UPDES program requires permits for the discharge of pollutants from any point source, including discharges from drinking water treatment plants.

In June of 1998, a general UPDES permit was issued for all drinking water plants in Utah. This permit is effective for five years, and contains a number of provisions. Drinking water plants are not permitted to have any discharges to the waters of the state, except as allowed by the provisions of the permit. The only allowable discharges for a drinking water plant under this permit include:

- Discharges from emergency overflow systems only in the event necessary to protect the plant and equipment
- Routine, excess, and presedimentation flows that have had no chemicals added

Drinking water plants are required to contain all recycle, such as spent filter backwash water, and direct discharge of these residuals is not permitted. The discharge from dewatering of water treatment chemical sludges is specifically forbidden by this permit.

Indirect discharge to sanitary sewer. Utah does not set state-wide limits for indirect discharges to surface waters through sanitary sewers. TBLLs for specific contaminants may be established by local sewer authorities or POTWs that administer pretreatment programs. Regulations provide that no pretreatment permit be issued for any indirect discharge from an industrial user that does not assure compliance with applicable pretreatment standards or requirements and that otherwise interferes with, passes through, or is incompatible with a POTW's treatment process, including contamination of sewage sludge.

Drinking water treatment residuals are required under a general Utah permit (UPDES) to be contained on site and recycled back to the head of the treatment plant. Therefore, no drinking water treatment plant in Utah has requested that its residuals be accepted by a POTW through the sanitary sewer. Because of the UPDES permit, this practice would not be allowed. In the past, other industrial users, such as gravel pit operations, have applied for indirect discharge agreements and have been allowed to discharge their wastes with a TBLL limit of 45 µg/L arsenic in one city.

Underground injection control. Utah administers the UIC program within the state, except for Indian lands. EPA Region 8 administers the program on Indian lands, and has oversight of the Utah program. Consistent with Federal UIC requirements, Utah prohibits underground injection that is not conducted with a permit and prohibits activity that "allows the movement of fluid containing any contaminants into underground sources of drinking water, if the presence of that contaminant may cause a violation of any primary drinking water standard or may otherwise adversely affect the public."

Solid Residuals

Landfilling and beneficial reuse. Utah has no specific regulations for drinking water treatment residuals for land application or landfilling of residuals that have passed the TCLP test for arsenic. In general, as discussed previously in the Federal regulation section, the RCRA requirements of best management practices are required, including containment (no runoff), dewatering, and disposal in a way that does not create a nuisance.

Colorado

Liquid Residuals

Direct discharge to surface waters. The State of Colorado administers the CWA's NPDES program through its Water Quality Control Division of the Department of Public Health and Environment's Colorado Discharge Permit System. A permit application, specifically for water plant discharges, is required for any drinking water treatment process stream that is discharged to a waterway. The conditions of the permit are specified on a case-by-case basis.

Indirect discharge to sanitary sewer. In Colorado, the Industrial Pretreatment program is administered by the EPA in Region 8, with coordination with the Colorado Water Quality staff. The POTWs are required to determine TBLLs for toxic materials coming into their plants. In general, the POTWs do not develop pretreatment standards on an industry by industry basis, but rather base it on the total amount of toxic material in their influents. Approximately 28 POTWs in Colorado have pretreatment standards in place for their facilities, and some of these POTWs accept residuals from drinking water plants. The TBLLs established for arsenic at most POTWs are generally less than 50 μg/L. In the future, it is expected that the TBLLs will be revised to limit toxic materials in terms of the total loading (pounds per day) instead of the current measure of mg/L, since this measure is of more significance for wastewater plants.

The limiting factor for the acceptance of drinking water residuals by a POTW is typically the level of suspended solids in the material. In general, arsenic has not been considered to be particularly toxic to wastewater plant operations, but additional arsenic in the WWTP influent could impact the final biosolids disposal method. Increased influent arsenic levels could also impact the quality of the effluent water from the POTW if breakthrough occurs, thereby violating the NPDES permit arsenic limit.

Underground injection control. In Colorado, the EPA in Region 8 has jurisdiction over Class I and V well permitting. In general, a water treatment residual with elevated arsenic could be disposed of in either type of well, if EPA would grant a permit. In granting similar permits in the past, the toxic pollutant levels in the waste were not allowed to exceed the respective maximum contaminant level (MCL) set for that toxic compound. Further, the facility had to demonstrate that no movement would be made from the well into an aquifer for 100 years.

Current federal requirements prohibit any injection activity that may endanger underground sources of drinking water. Under these regulations, the EPA and the states have the authority to bring wells that are endangering underground sources of drinking water into compliance, and take enforcement and closure actions where necessary.

Solid Residuals

Landfilling. The State of Colorado administers the provisions of RCRA for the regulation of landfill disposal of drinking water treatment residuals. In the case of drinking water treatment residuals suspected of high levels of toxic contaminants such as arsenic, stringent monitoring of the site would be required to confirm that no degradation of the groundwater occurs. In order to determine if a material is toxic, and so required to be handled under RCRA Subtitle C, the State of Colorado uses the TCLP as well as the land application limits set forth in 40 CFR 503(b).

Beneficial reuse. The Colorado Department of Health allows for the beneficial land application of drinking water treatment residuals through a "beneficial use certification" on a case-by-case basis. The department has established specific regulations pertaining to the beneficial use of water treatment residuals on land. These regulations require development of an approved beneficial use plan that:

- Identifies where the material will be used
- Includes residuals analyses
- Identifies the types of crops to be grown and the application rates

Parameters to be analyzed in the permit, and then annually if the permit is granted, are shown in Table 2.2 and report arsenic in mg/kg. Application to land where root crops and low growing fruit and vegetables are grown is prohibited if the crops are intended for human consumption. Co-disposal of water treatment residuals with wastewater biosolids is permitted only if the combined material complies with all applicable requirements of the Colorado Biosolids Regulations. These state regulations include the land application limits of the Clean Water Act's 503(b) regulations discussed earlier. Typically, the application rate of the mixed residuals is limited by the phosphorous fixing capability of the material.

Table 2.2

Analyses and reporting units for Colorado beneficial use permit

Parameter	Units*	Parameter	Units*
Total solids	Percent	Total Cr	mg/kg
pH	Standard units	Total Cu	mg/kg
Organic-N	Percent	Total Fe	mg/kg
Total ammonia-N	Percent	Total Pb	mg/kg
Nitrate-N	Percent	Total Hg	mg/kg
Total P	Percent	Total Mb	mg/kg
Total K	Percent	Total Ni	mg/kg
Total Al	mg/kg	Total Se	mg/kg
Total As	mg/kg	Total Zn	mg/kg
Total Cd	mg/kg	Total alpha activity	pci/g†

*All results expressed in dry weight basis for a composited sample
†Picocuries per gram

Montana

Liquid Residuals

Direct discharge to surface waters. Montana has set up its own pollutant discharge program called the Montana Pollutant Discharge Elimination System (MPDES) with the goal to control point source discharges in order to protect the water quality of receiving streams. The standards for water quality that are to be met for various uses of the streams are described in the state Water Quality Standards (WQS). All point source dischargers are required to obtain a MPDES permit in which the effluent limitations are based upon preservation of WQS, and other specified contaminants must also be treated to a technology-based minimal level before being discharged. WQS requirements for pollutant levels are calculated at the average design WTP residuals flow and the seven-day, ten-year low stream flow (7Q10) in the receiving stream.

Within the code of the MPDES, Montana has established Nondegradation Rules that are applicable to new or increased sources of pollution as a part of the WQS. These rules prohibit discharge of new or increased sources of pollution, suspected toxins, and deleterious materials to state waters unless it has been proven to the Department of Environmental Quality (DEQ) that economic or social development warrants a change in the regulation.

In addition to heavy metals, nutrients and toxic organic pollutants are limited under the Nondegradation Rules. These pollutants may also be limited under the WQS in existing permits. However, the WQS levels are calculated to meet less than chronic toxicity levels at the 7Q10 and nondegradataion limits in new or enlarged point source discharges are set at baseline instream concentrations as well as "trigger level" amount that defines the "significance" threshold.

Also, within the MPDES are bylaws allowing mixing zones. A mixing zone is defined as an area of surface or groundwater where the initial dilution of the pollutant is discharged and the contaminant concentration levels are knowingly exceeded. There are numerous restrictions on the size, location, and changes that are allowed within a mixing zone, and only the DEQ may grant a mixing zone. Arsenic is one of many pollutants that is permitted to be discharged in a mixing zone in high concentrations, however, there are no WTPs currently using mixing zones to discharge arsenic-laden residuals. All of the existing mixing zone permits that allow large doses of arsenic have been granted to industrial users.

Indirect discharge to sanitary sewer. In Montana, the Region 8 EPA office has authority for the Industrial Pretreatment Program. The program is administered as described in the previous section, with POTWs required to determine TBLLs for toxic materials coming into their plants. In general, the POTWs do not develop pretreatment standards on an industry by industry basis, but rather base it on the total amount of toxic material in the influent of their plant. Only one POTW has pretreatment program standards in place for its facility.

Underground injection control. In Montana, the EPA in Region 8 has jurisdiction over Class I and V well permitting. In general, a water treatment residual with elevated arsenic could be disposed of in either type of well, if EPA would grant a permit. In granting similar permits in the past, the toxic pollutant levels in the waste were not allowed to exceed the respective maximum contaminant level (MCL) set for that toxin. Further, the facility had to demonstrate that no movement would be made from the well into an aquifer for 100 years.

Current federal requirements prohibit any injection activity that may endanger underground sources of drinking water. Under these regulations, the EPA and the states have the authority to bring wells that are endangering underground sources of drinking water into compliance and take enforcement and closure actions where necessary.

The underground injection regulations for the State of Montana are consistent with federal regulations as delegated and administered by the EPA in the Underground Injection Control Program (UIC). The UIC regulations are in compliance with the Federal Safe Water Drinking Act and address procedures for revision of state programs at 40 CFR 145.32. Currently, there are over 1,400 injection wells that are regulated and inspected for compliance and mechanical integrity semi-annually. There are no additional state regulations that address the issue of arsenic disposal.

Lagoon disposal. All waste or product holding facilities such as lagoons, leaching pads, and tailing ponds must be designed, constructed, operated, and maintained to prevent discharge, seepage, drainage, infiltration, or flow that may result in the pollution of state waters. All plans for facilities used in ore processing must be submitted to the DEQ for review at least 180 days prior to the commencement of operations. Adequate provisions are required to seal the sludge lagoon bottoms and embankments to prevent leaching into adjacent soils or groundwater. For ultimate disposal, lagoons are not allowed.

Solid Residuals

Landfilling. Montana uses the Federal RCRA program to manage its hazardous waste landfills in addition to the Montana Hazardous Waste Act. This act establishes a management control system for permitting the waste through its life-cycle, while simplifying the language of the regulations.

Beneficial reuse. Montana encourages the beneficial use of sludge in all cases. The DEQ must be contacted for specific design and approval requirements governing land application of municipal sludges. Montana has established its sludge characteristic monitoring program and requires arsenic to be measured quarterly. Additional operating criteria may be obtained from applicable Federal regulations.

Incineration. Montana also practices burning of hazardous wastes in boilers and industrial furnaces for both the energy recovery or destruction and for the material recovery. Guidelines for the feeding rate of arsenic and other potentially dangerous metals is described in Montana's regulations such that the industrial furnace feed stocks shall not exceed specific values derived from the screening limits. These feed rate screening limits for carcinogenic metals are based upon an hourly rolling average or an average from a designated period.

Due to the toxic nature of the emissions from incineration, there are additional restrictions on the allowable height of the stack. The height is determined using a formula that sums the actual physical stack height and the plume rise and subtracts from this the terrain rise within 5 km of the stack. The emissions rate of carcinogenic metals like arsenic is not to exceed values derived from the screening limits specified.

Kansas

Liquid Residuals

Direct discharge to surface waters. The discharge of WTP residuals to Kansas surface waters is controlled via the issuance of discharge permits. These permits are referred to as Kansas Water Pollution Control Permits or NPDES permits. These permits are issued jointly by Kansas Department of Health and Environment (KDHE) and the EPA.

Permit limits for the discharge of WTP residuals are based on meeting technology-based limits, water quality criteria, or on best professional judgement (BPJ). Limits are imposed to protect water quality and limit degradation to existing state water quality.

Water quality effluent limits (WQBELS) for industrial and municipal facilities and limitations are based on the total maximum daily load (TMDL) to receiving streams. Metals loading limits to surface waters are defined by Kansas regulation (K.A.R. 28e). The regulatory limits for disposal of WTP residuals containing arsenic are listed in Table 2.3.

Table 2.3

Kansas discharge limits for arsenic

Metal (μg/L)	Aquatic life		Agriculture		Public health	
	Acute	Chronic	Livestock	Irrigation	Food procurement	Domestic water supply
Total Arsenic	-	50	200	100	-	50
Arsenic (III)	379	50	-	-	-	-
Arsenic (IV)	850	48	-	-	-	-

- Limit not available

Indirect discharge to sanitary sewer. EPA Region 7 administers the Pretreatment Program for Kansas POTWs following the 40 CFR 403 guidelines. The Pretreatment program is intended to control pollutants that are incompatible, will interfere with, or will pass through POTWs. In addition, pretreatment requirements are intended to improve opportunities to recycle and reclaim domestic and industrial wastewaters and sludge and enable communities to comply with water quality standards in receiving streams. Components of pretreatment programs include an adequate ordinance, public education, industrial inspection, and a permit system to monitor industrial discharges.

Traditionally, the smaller POTWs (<5.0 mgd) with individual discharges are not required to establish local pretreatment programs. If non-domestic users in the community (such as foundries, battery manufactures, metal finishers, etc.) are discharging toxic pollutants that could pass through the POTW system untreated or interfere with its operations, the community may have to develop and implement a pretreatment program.

Underground injection control. The Federal Safe Drinking Water Act requires states to meet Federal Environmental Protection Agency standards for regulating underground injection of fluids. The Kansas Conservation Division has sole responsibility for implementing the Underground Injection Control Program for Class II (oil/gas) wells in Kansas. State statutes require that oil and gas operators have approval prior to fluid injection and that certain records and tests be made. Injection applications are technically reviewed to confirm that fresh and usable water, hydrocarbons, and correlative rights will be protected. Operators must report well operating statistics (i.e., injection pressures, rates, etc.), and wells must have mechanical integrity tests run regularly.

Solid Residuals

Landfilling. Kansas has imposed stringent regulations on landfilling of solid wastes. Landfill sites are required to monitor groundwater under and around the landfill for traces of arsenic. Monitoring wells are required to determine the concentration of contaminants at various locations surrounding the site. Arsenic concentrations should not exceed a maximum contaminant level of 0.05 mg/L at the uppermost aquifer at the designated point of compliance. The point of compliance must be a location on the facility's site that is 150 meters from the edge of the landfill and at least 15 meters from the property boundary.

Lagoon disposal. Kansas Statutes and Commission Regulations require permitting of surface ponds prior to use. Applications for surface ponds are reviewed by technical and field staff for pond construction information and location so that protection of fresh and usable water is confirmed prior to application approval. Statutes further require that spillage of fluids associated with oil and gas production be immediately reported to the appropriate Kansas Corporation Commission district field office and that all spills are cleaned up in a timely manner. Pits containing a discharge shall be emptied within 48 hours after the discharge occurs. The Commission also has the responsibility to have operators clean up pollution to soil and water resources from oil- and gas-related activities. In certain cases, operators are required to develop plans for clean up of contamination that are reviewed by technical staff for consistency with Commission regulations and objectives.

Beneficial reuse. The Kansas Department of Health and Environment currently does not have any specific guidelines governing the disposal of WTP coagulant residuals by land application or other beneficial use applications. Approval for these types of residuals disposal alternatives are regulated on a case-by-case basis. The State of Kansas does not encourage land applying WTP coagulant sludges, but instead prefers lagoon storage or dewatering and placement in a solid waste landfill. In order to lagoon coagulant residuals, a permit is required from the KDHE Bureau of Water.

Minnesota

Liquid Residuals

Direct discharge to surface waters. Minnesota is authorized by EPA Region 5 to administer the NPDES program through the Minnesota Pollution Control Agency (MPCA). The MPCA is the agency responsible for developing water quality standards and determining effluent limitations for surface water discharges in the state. Both numeric and narrative standards are developed for the various beneficial use designations prescribed in Minnesota rules. The primary surface water rules pertaining to arsenic are Chapter 7050, which applies statewide, and Chapter 7052, which applies to the Lake Superior Basin. The numeric water quality standards for protection of cold, cool, and warm water aquatic life, or drinking water classifications for both rules are listed in Tables 2.4 and 2.5.

Table 2.4

Arsenic surface water standards (µg/L) in Chapter 7050 of MPCA regulations

Classification in Chapter 7050	Chronic standard (cs)	Acute aquatic life maximum standard (ms)	Aquatic life final acute value (FAV)
2A cold water	2*	360	720
2Bd cool water	2*	360	720
2B cool water	53†	360	720
2C warm water	53†	360	720
2D wetlands	53†	360	720

*Reflects protection as source of drinking water
†Reflects protection of human health fish consumption and incidental ingestion of water

Table 2.5

Arsenic surface water standards (μg/L) in Chapter 7052 (Lake Superior Basin) regulations of MPCA

Classification in Chapter 7050	Chronic standard (cs)	Acute aquatic life maximum standard (ms)	Aquatic life final acute value (FAV)
2A cold water - Lake Superior	2*	340	680
2A cold water - inland	2*	340	680
2Bd cool water	2*	340	680
2B cool water	53†	340	680
2C warm water	53†	340	680
2D wetlands	53†	340	680

*Reflects protection as source of drinking water as derived from Chapter 7050
†Reflects protection of human health fish consumption and incidental ingestion of water as derived from Chapter 7050

The Final Acute Value (FAV) serves as an upper concentration limit applied as a daily maximum effluent limitation. As the theoretical water quality based effluent limitations for a discharge increase with dilution, the FAV serves as the maximum limit allowable (based on requirements in Chapter 7050). In Chapter 7050, the FAV serves as a default cap unless the discharger supplies information demonstrating that an acute mixing zone would allow for a higher discharge concentration.

The exposure duration associated with acute standards is a 1-day average. A 4-day duration is associated with aquatic life chronic standards, and a 30-day duration applies to human health chronic standards.

Chapter 7050 contains narrative provisions applicable to arsenic relating to general nuisance conditions, which could be utilized where there is concern for excessive sediment build-up. Chapter 7052 contains similar general language for mixing zone provisions.

Indirect discharge to sanitary sewer. In order to discharge WTP residuals to POTW, a Minnesota Pollution Control Agency (MPCA) State Disposal System (SDS) Pretreatment Permit is required. The State SDS approval requires the waste generator to list the POTW(s) that would receive the residuals and the specific quality of the waste stream. For WTP residuals containing arsenic, the concentration of arsenic and residuals quantity would be required. Permit approval is based on the tolerance of the POTW to handle additional arsenic, while still meeting their own biosolids and treated effluent limits. Therefore, the SDS pretreatment permits are authorized on a case-by-case basis by MPCA and the local POTW involved.

Underground injection control. EPA Region 5 has retained primacy for the UIC program in this state, so the Federal regulations detailed in the previous section (Federal Regulatory Survey) are used. In general, water treatment process wastes could be disposed of in either a Class I hazardous deep well, or a nonhazardous Class 5 shallow well if the EPA would approve the permit. Minnesota has very few Class 5 wells permitted currently.

Solid Residuals

Landfilling. The MPCA regulations governing landfill disposal of non-hazardous solid or industrial wastes are defined in the Minnesota Statutes 2000, 115A and 115B. Hazardous characterization is determined by the TCLP test using the limits outlined in 40 CFR 258 landfill regulations. In order to landfill WTP residuals, they must contain no free liquids as determined by the Paint Filter Test. WTP residuals that fail the TCLP test for arsenic or other regulated contaminants would be re-classified as a hazardous waste and would require disposal in a Class I hazardous waste landfill facility.

Beneficial reuse. WTPs that wish to beneficially reuse their residuals via land application or other forms of land disposal must first obtain an MPCA Industrial By-Product Land Application Permit (MNG960000). This permit is part of the State NPDES and SDS Permitting Program. Each industrial waste is reviewed on a case-by-case basis to determine acceptance for beneficial reuse.

The General Permit criteria are identified by MNG960000 Part I.2.1 through I.2.2, and require that the waste be non-hazardous based on TCLP analysis and in terms of total contaminant concentrations. Table 2.6 provides the total contaminant concentrations that are acceptable in industrial by-products.

Table 2.6

Concentration limits for industrial by-products

Analyte	mg/kg (dry wt. basis)
Total As	41
Total Cd	39
Total Cu	1,500
Total Pb	300
Total Hg	5
Total Ni	420
Total Se	100
Total Zn	2,800
Total dioxin equivalents	10 ppt
Total polychlorinated biphenyls (PCBs)	6

If land application is utilized, then the annual loading rate limit allowed under the MPCA General Permit for arsenic application is 1.8 lb/acre/yr.

Wisconsin

Liquid Residuals

Direct discharge to surface waters. The State of Wisconsin's Department of Natural Resources (DNR) has the administrative authority to carry out the NPDES in Wisconsin, with the EPA in Region 5 having oversight responsibilities. The DNR regulates municipalities, industrial facilities and significant animal waste operations discharging wastewater to surface waters or groundwater of the State of Wisconsin through the Wisconsin Pollutant Discharge Elimination System (WPDES) Permit Program. No person or entity may legally discharge to waters of the state without a permit issued under this authority. This permit system is patterned after the Federal

NPDES program administered by the EPA. All permits issued under the WPDES permit program are either specific permits or general permits. Specific permits are issued to individual facilities. General permits are issued statewide to cover facilities with similar discharges. The DNR makes a determination on whether a particular facility is appropriately covered by a general or specific permit. With respect to EPA's oversight, all specific WPDES permits issued by the DNR are classified as major or minor WPDES permits. Major WPDES permits are subject to EPA oversight and concurrence from EPA must be received prior to finalizing any permit action. "Point source" dischargers (i.e., facilities discharging wastewater to surface water from a specific point such as from the end of a pipe) must meet either the Federal minimum requirements for secondary treatment in the case of a municipality, and technology-based categorical (base level) limits for industries or discharge levels necessary to achieve water quality standards, whichever is more stringent. Land disposal systems also receive permits with limits established to protect groundwater. In addition, WPDES permits address safe limits for land application of municipal and industrial wastewater biosolids. Where necessary to achieve effluent limits, a WPDES permit includes a compliance schedule for making improvements to the treatment system.

In 1979, Chapter 147.023 (since renamed and now known as Ch. 283.35, Wisconsin Statutes) was created to authorize the DNR to issue WPDES General Discharge Permits. A General Discharge Permit is designed to cover groups of facilities or industries with similar types of wastewater discharges to surface water and/or groundwater. There are currently 15 active industrial general permits and one active municipal general permit. A general permit for Potable Water Treatment and conditioning (WI-0046540-3), would cover any residual stream from an arsenic removal process.

This permit, reissued in May of 2001, is intended to cover iron filter, lime softening, alum coagulation, granular media filtration, and reverse osmosis facilities where backwash, regeneration, and rinse water are discharged to surface waters or land treatment systems. It contains specific water quality effluent-based limits for a number of contaminants, but not for arsenic. The DNR is reviewing this permit since the revised arsenic MCL has been promulgated. Presently, the Wisconsin water quality standard for chronic exposure to As(III) for the protection of aquatic life is set at 148 µg/L.

Indirect discharge to sanitary sewer. The State of Wisconsin has primacy for the Industrial Waste Pretreatment Program with EPA oversight. Although the state has only 20 plants that use surface waters, these are mostly large plants, and most of the liquid residuals are sent to the sanitary sewer. The state has indicated that they may ban the practice of recycling spent filter backwash water to the head of the plant after the promulgation of the Filter Backwash Rule (FBRR). If this occurs, then the quantity of this material being sent to the sewer will increase. Presently, the local limits for sewer discharge are set by the POTWs, and an informal survey of the POTWs did not find any local limits set for arsenic.

Underground injection control. The State of Wisconsin has authorization for the Underground Injection Control Program of the SDWA with EPA Region 5 oversight. Disposal of residuals from drinking water treatment processes would not be possible under this program, as the state has banned the permitting of any new wells of any class.

Lagoon storage. The Wisconsin Department of Natural Resources has identified storage lagoons, or evaporation ponds, for the storage and ultimate (at least partial) disposal of liquid residuals through percolation into the groundwater. This method of storage must meet the SDWA requirements discussed earlier as well as the conditions of a permit issued by the EPA Office of Groundwater for discharge to a groundwater source.

Solid Residuals

Landfilling. Wisconsin was the first state to receive approval of its solid waste program by the USEPA. The Wisconsin Department of Natural Resources (DNR) regulates the municipal and industrial waste landfills in the state. There are currently 44 municipal solid waste landfills and 41 industrial waste landfills. There are currently no hazardous waste landfills operating in the state. Landfills are regulated by the DNR NR500 administrative code series.

In Wisconsin, WTP solids are considered an industrial waste, which if determined to be non-hazardous by the TCLP test (40 CFR 258), could be disposed of in solid waste or industrial waste landfills. The residuals must also be dewatered such that no free liquids are present.

Beneficial reuse. The regulations governing the land spreading of solid waste in Wisconsin are outlined in DNR Chapter NR 518. The goal of this regulation is to confirm that environmentally acceptable solid waste management procedures are used for land applying wastes. The state allows land spreading of industrial sludges as long as the following criteria are met in accordance with NR 518.06.

These are as follows:

- Residuals are determined to be a soil conditioner or fertilizer by WDNR. Demonstration testing may be required prior to acceptance
- Residuals must be generated and disposed by approved facilities under WDNR Chapter NR 214
- Residuals are not repeatedly applied such that hazardous substances accumulate in the soil, vegetation, or cause detrimental impacts on groundwater quality

Prior to approval for land spreading, a detailed description of the waste material must be submitted to WDNR for approval. The waste characterization requirements (listed in NR 518.06) are as follows:

- List of waste sources
- Pretreatment used
- Volume for disposal
- Physical and chemical analysis
- Waste use determination
- Operating and monitoring plan

The WDNR provides approval for land spreading on a case-by-case basis.

Indiana

Liquid Residuals

Direct Discharge to Surface Waters. Indiana's Department of Environmental Management's (IDEM) Office of Water Quality (OWQ) implements and enforces the Federal Water Pollution Control Act (as amended), also referred to as the Clean Water Act. With oversight from EPA Region 5, IDEM's OWQ Wastewater Permitting Branch assumed responsibility for this permit program in 1975. The Act prohibits the discharge of a pollutant into the "waters of the United States" as a point source discharge without a Federal NPDES permit.

Water Quality Based Effluent Limits (WQBELs) are designed such that the quality of any wastewater being discharged will not cause any acute or chronic adverse effects to occur in the receiving water body. The WQBELs are also designed such that all beneficial uses of the receiving water body are protected and maintained. Indiana water quality criteria are established through the rule-making process for the pollutants found in wastewater. Indiana water quality standards use the water quality criteria to establish the maximum concentration of pollutants that may be present in a water body and will not cause acute or chronic toxicity to aquatic life, terrestrial life (wildlife which use the water body), or humans (who consume the water or fish from the water body).

Mathematical models are used to calculate the WQBELs for wastewater discharges that contain toxic or oxygen demanding pollutants. The models consider the following factors when calculating the WQBELs: the design flow rate of the discharger, the background concentration of the pollutants in the receiving water body, the average minimum seven consecutive day low flow that occurs once in ten years (7Q10) at the location of the discharge, and other hydrologic characteristics of the receiving water body.

Indirect discharges to sanitary sewer. The EPA Region 5 presently has the authority to administer the Industrial Pretreatment Program in Indiana. As such, the EPA has technical approval authority for all permits issued under this program, but the individual POTWs in Indiana are responsible for the development of these programs. Wastewater from drinking water treatment processes sent to a POTW through sewer lines must first be treated and receive an Industrial Wastewater Pretreatment Permit (IWPP). IWPPs are for industrial process wastewater, which is treated to remove contaminants prior to discharge into a municipal wastewater collection system. Treatment is similar to that associated with NPDES permits, but the effluent is discharged into a municipal sewer rather than directly into a stream or other body of water. As a result, this wastewater receives further treatment at the municipal facility prior to being discharged to waters of the state.

Currently 45 Indiana municipalities have EPA-delegated pretreatment programs in place, under which they regulate industrial discharges to their POTWs. IWPP effluent standards for arsenic in drinking water processes in Indiana are derived in one of two ways:

- POTW calculations that derive effluent limits by back-calculating the amount of pollutant loads available to industry, taking into account the available additional capacity of the POTW before its capabilities are exceeded, or
- Local limits established in a Pretreatment Program Ordinance

Most of these programs have included local limits for arsenic. For example, the local limit for the Kokomo, Indiana water treatment facility was set by permit at 5 mg/L. The POTW is required to sample its influent for this contaminant once yearly, and the industrial dischargers, including any water treatment plants discharging to the sewer, must sample their effluents at least twice per year.

Underground injection control. The State of Indiana has not sought primacy of the portion of the SDWA directed to the UIC Program other than for Class II wells, which are administered by the Indiana Department of Natural Resources. The UIC Program for Class I, III, and IV wells is currently administered by EPA Region 5. The UIC Program serves functionally as the NPDES program for groundwater discharges.

Solid Residuals

Landfilling. In the past, WTP residuals were classified by the State of Indiana as "special wastes," however, effective July 2000, these wastes are now referred to as "industrial wastes." Industrial wastes are defined as solid waste from a non-residential source that is not hazardous waste, municipal waste, construction or demolition waste, or infectious waste.

The landfill regulations for Indiana are provided by the IDEM under IDEM identification number WASTE-0001-NPD. Prior to disposal, the waste material must be classified for hazard potential. Hazardous determination is outlined by RCRA criteria (450 CFR 261) and includes TCLP toxicity, reactivity, corrosivity, and ignitability. Landfills also require the wastes be dewatered such that it contains no free liquids.

Beneficial reuse. Facilities in the State of Indiana that generate non-hazardous industrial wastes have the option of land applying selected industrial wastes to avoid tipping fees associated with landfilling. The beneficial uses possible include land application on agricultural fields, reclamation projects, and land application at other sites. To land apply, a facility must have a permit from IDEM's land application program. Permit acceptance is based on the quality, quantity, and type of disposal requested. IDEM generally follows the 40 CFR 503 guidelines for both biosolids and industrial waste land application. Therefore, the specific criteria for disposal of residuals containing arsenic is 75 mg/kg ceiling concentration and 41 mg/kg pollutant concentration.

IDEM issues two types of permits for land application of industrial wastes as follows:

- Non Site-Specific - use on any agricultural land
- Site-Specific - use on selected sites

IDEM also uses a hybrid permit that allows both non- and site-specific applications. Facilities that wish to market or distribute residuals must obtain a Marketing and Distribution Permit from IDEM.

Texas

Liquid Residuals

Direct discharge to surface waters. The Texas Natural Resource Conservation Commission (TNRCC) received permission to begin discharging wastes under the guidelines of the Federal NPDES Program from the EPA on September 14, 1998. The Texas version is referred to as a Texas Pollutant Discharge Elimination System (TPDES) permit, with guidelines listed in the Texas

Administrative Code (TAC). The guidelines allow for discharge of wastes from industrial and domestic wastewater treatment works, storm runoff from industrial activities, construction sites, city storm sewer systems, and wastes sludge from sewage treatment plants. WTP residuals are considered to be an industrial waste. In order to discharge any of these wastes, the operator must be permitted and strictly follow all guidelines set up by the NPDES and TPDES.

Discharge of residuals to surface waters are regulated based on the toxic metals concentrations of the liquid waste. TAC 307.6 lists the criteria in water for toxic materials. The arsenic limits from this regulation are summarized below in Table 2.7.

Table 2.7

Texas surface water quality standards for aquatic life

Parameter	Freshwater acute criteria (μg/L)	Freshwater chronic criteria (μg/L)
As	360W	190W

W - EPA conversion factor multiplied by water effects ratio

The arsenic criteria for human health protection is 50 μg/L, based on the MCL specified in 30 TAC 290 relating to water hygiene. Also, the TNRCC general regulations state under 319.22 and 319.23 that the allowable concentration of arsenic for discharge to inland waters should not exceed an average of 100 μg/L.

Indirect discharge to sanitary sewer. The State of Texas has an approved State pretreatment program for disposal of liquid wastes to POTWs. TNRCC follows the EPA guidelines for the pretreatment program as provided in the Federal 40 CFR 403 regulations.

Underground injection control. The State of Texas regulates underground injection of non-hazardous liquid wastes through 30 TAC 331, which implements the provisions of the Injection Well Act, Texas Water Code Chapter 27. The State assumes primacy for regulation of all wells within this jurisdiction. Liquid discharges to injection wells are regulated on a case-by-case basis, depending on the well location and the quality of waste disposal.

Solid Residuals

Landfilling. Landfill disposal of WTP residuals in the State of Texas requires a TNRCC permit in accordance with 30 TAC Chapter 305 (Consolidated Permits). Waste disposal must be consistent with the Solid Waste Disposal Act criteria (Texas Civil Statutes, Article 4477-7) that includes proper collection, handling, storage, processing, and disposal. Texas Landfill requirements generally follow the EPA 40 CFR 257 regulations, and require that the material be non-hazardous as defined by EPA regulation 40 CFR 261 and classified according to 30 TAC 335 Subchapter R. Under this regulation, solid wastes are classified as hazardous, Class I, Class II, or Class III.

Beneficial reuse. In order to land apply WTP residuals, an application must be submitted to TNRCC's permitting Section of the Watershed Management Division. The beneficial use disposal activity must meet the requirements of EPA 40 CFR 257. According to 30 TAC 312.63, the arsenic concentration for land application should not exceed 73 mg/kg. Utilities that land apply WTP residuals must submit a status report to the executive director each year.

New Hampshire

Liquid Residuals

Direct discharge to surface waters. In the State of New Hampshire, EPA Region 1 administers the Federal NPDES program for discharge of wastes to surface water. The regulations are implemented in accordance with Section 402 of the Clean Water Act. Facilities that desire to discharge WTP residuals must obtain a Federal NPDES permit.

Indirect discharge to sanitary sewer. The State of New Hampshire does not have an approved state pretreatment program and, therefore, its program is administered by the EPA Region 1 Office. Part ENV WS904 of the NHDES guidelines provides the standards for pretreatment of industrial wastewaters to POTW. These guidelines were directly developed from the EPA 40 CFR 403 General Pretreatment Regulations for Existing and New Pollution. The direct discharge of WTP residuals is dependant on the local limits set by municipal sewer ordinances, which numerically limit the amount of each specific pollutant that can be tolerated by the POTW.

Underground injection control. NHDES has implemented a Groundwater Discharge Permitting and Registration Program that is concerned with the proper treatment and disposal of wastewaters onto or into the ground. All wastewater discharges to the ground must be registered, or in some cases permitted by the NHDES Water Division. These discharges also include lagoons or land application of liquid WTP residuals. This program implements Federal regulations pertaining to underground injection wells. The groundwater discharge permit is issued by NHDES under RSA 485-A:13 and ENV-WS1500. The NHDES sets forth an ambient groundwater quality standard for a number of contaminants including arsenic. The NHDES water quality standard for arsenic is 50 μg/L.

Solid Residuals

Landfilling. The State of New Hampshire has received Federal approval to administer the 40 CFR 258 landfill regulations. The NHDES developed Solid Waste Rules as landfill regulatory guidelines. Landfilling WTP residuals in New Hampshire requires the same physical and chemical characteristics as determined by the 40 CFR 258 regulations, including a non-hazardous determination and no free liquids.

Beneficial reuse. The NHDES Sludge Management Rules (Ch. ENV-WS 800) provide the standards and criteria for a permit system to manage disposal of sludge by beneficial use recycling. The NHDES outlines specific requirements necessary to obtain a permit including:

- Site facility permit
- Sludge quality certification
- Sludge hauling permit
- Disposal site requirements
- Land application standards

Section ENV-WS 806.08 outlines the specific metals concentrations that are regulated for land application of sludge. The cumulative loading rate for arsenic is 9 lb/acre (10kg/ha), and the total arsenic concentration should not exceed 32 mg/kg and 19 mg/kg for low-metals sludge application. The land application program must be conducted using continued testing, monitoring, and reporting to NHDES.

Louisiana

Liquid Residuals

Direct discharge to surface waters. The Louisiana Department of Environmental Quality (LADEQ) regulates the environmental discharge of liquid wastes using the Federal NPDES program 40 CFR Part 503 as a guideline. The LADEQ program requires permits for the discharge of point source pollutants into receiving waters.

The LADEQ Title 33 regulations Chapter 11 provides the State "Surface Water Quality Standards" for discharge into receiving streams. The DEQ has set numerical criteria for specific toxic metals. The arsenic criteria are shown in Table 2.8.

Table 2.8

Numerical criteria for arsenic in LA receiving streams

	Aquatic life protection				
	Freshwater		Marine		
Metal	Acute	Chronic	Acute	Chronic	Drinking water supply
Arsenic (µg/L)	339.8	150	69.0	36.0	50.0

Indirect discharge to sanitary sewer. The State of Louisiana has an EPA-approved State pretreatment program for regulating non-domestic wastewater discharge to POTWs. These regulatory statutes are outlined in LAC 33:IX.2705 of the Louisiana DEQ Code of regulations. Local limits are developed by individual POTWs to minimize treatment interference.

Underground injection control. The Louisiana Department of Natural Resources Office of Conservation's Injection and Mining Division is responsible for administering the EPA- approved Underground Injection Control (UIC) program.

Solid Residuals

Landfilling. The Louisiana Solid Waste Regulations, outlined in LAC 33: Part VII Subpart 1, establish standards for governing the storage, collection, recovery, reuse, and disposal of solid waste. In Louisiana, the sludges resulting from the treatment of water (public or private WTPs) are not subject to the permitting process or disposal standards outlined in these regulations. These WTP residuals, however, must be deemed non-hazardous based on the TCLP analysis in order to be disposed of in municipal solid waste landfills. The WTP residuals must also be dewatered to contain no free liquids prior to landfill acceptance.

Beneficial reuse. In Louisiana, a utility must obtain a beneficial use permit prior to disposal of WTP solids by land application or other methods. These guidelines are provided by LAC 33: VII Chapters 1, 3, 5, 7 and 9. No permit can be issued unless the applicant supplies written documentation from an independent third party such as the Cooperative Extension, Department of Agriculture, or other stating that the proposed application is a legitimate beneficial use of the material. The operational plan should include a characterization of the waste material (including physical, chemical, and biological characteristics), minimum quantities disposed of per year, application rates, etc., as outlined by Section 1107 of the Beneficial Use Facility Regulations. The DEQ Solid Waste Division permits beneficial use on a case-by-case basis.

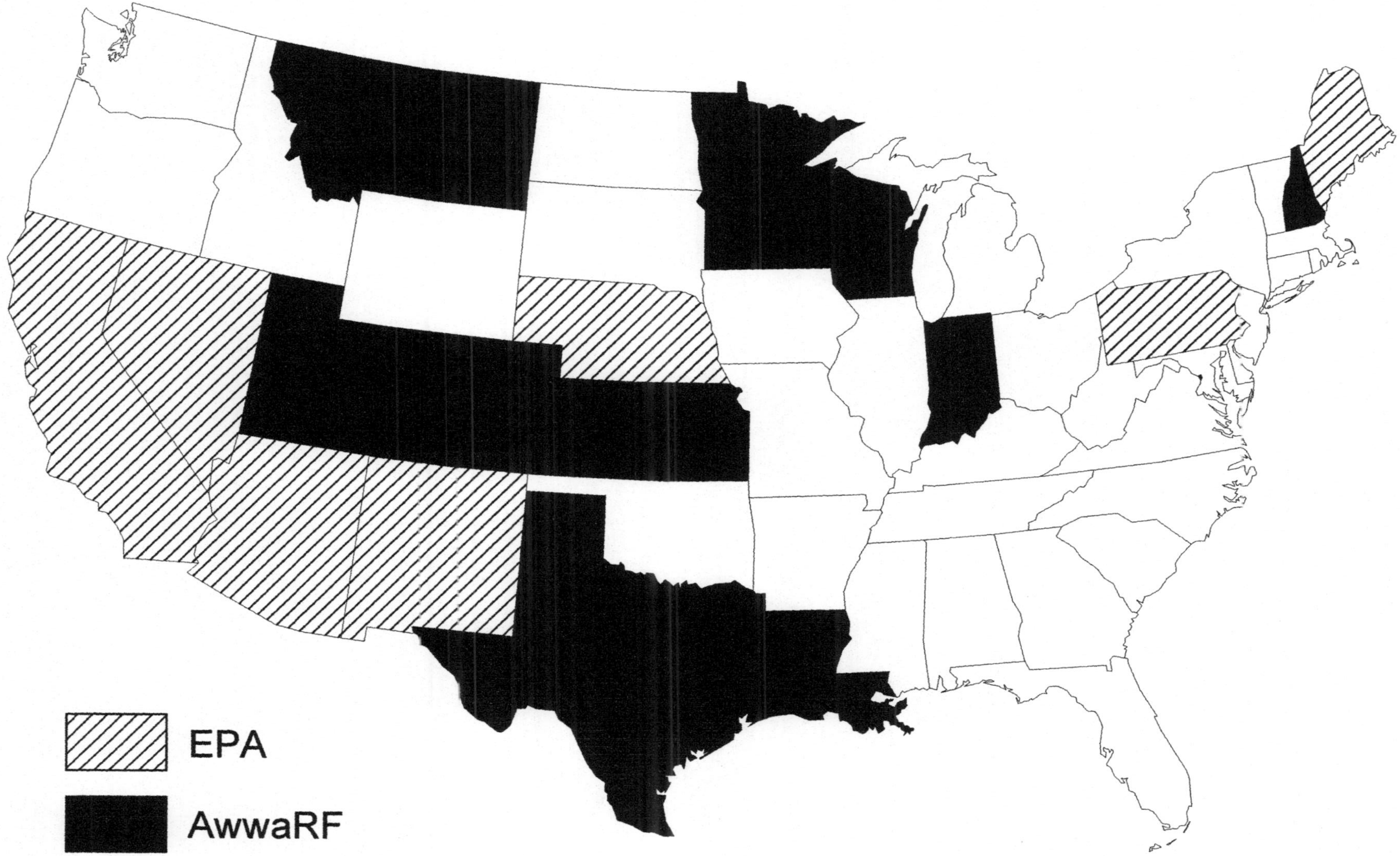

Figure 2.1 Map of states surveyed by both the EPA and AwwaRF projects

CHAPTER 3
CHEMICAL CHARACTERIZATION

UTILITY SAMPLING

Collection of the residuals samples were performed by utility staff using instructions provided by the project team (See Appendix C). Written sampling guidelines and plastic containers were shipped to a number of utilities from which a total of eight different residuals samples were obtained. All of the samples received included settled residuals, and six of these utilities also provided spent filter backwash (SFBW) samples. Each WTP was instructed to ship a two-gal container of settled residuals and a 20-gal sample of SFBW. The samples received are summarized in Table 3.1. A process schematic for each of these utilities, except for Heath, OH is provided in Appendix A.

Table 3.1

Residuals samples collected from participating utilities

Utility name	Samples type	Coagulant type
Louisiana Water Company (New Iberia), LA	Settled residual/SFBW	Lime softening
City of Great Falls, MT	Settled residual/SFBW	Alum
City of Helena, MT	Settled residual/SFBW	Alum
Billings Dept. of Public Utilities (Billings), MT	Settled residual/SFBW	Ferric
Lockwood Water Users Association (Lockwood), MT	Settled residual/SFBW	Ferric
Los Angeles Dept. of Water & Power (LADWP), CA	Settled residual/SFBW	Ferric
Indiana-American Water Company, IN	Settled residual	Ferric
Heath, OH	Settled residual	Lime softening

The original test plan called for collection of eight settled residuals samples (two lime, two alum, and four ferric) and six SFBW samples. The second lime softening facility (Heath, OH) was identified at the very end of the study, and residuals samples were obtained for characterization only.

SETTLED RESIDUALS CHARACTERIZATION

The settled residuals samples were delivered to Virginia Tech for analyses. Each sample was divided into three portions in order to generate the different residual streams required for analysis:

- Combined residuals (completely mixed as delivered)
- Secondary residuals (separated into liquid and solid portions for analysis) These residuals were gravity settled to separate the solids from the liquid portion, and then the solids were either air or oven dried prior to analysis

The untreated (combined) residuals for each utility were analyzed to determine the solids concentration and pH, while the two secondary residual streams (liquid and solids) were analyzed for the following parameters:

- Total arsenic
- Total aluminum
- Total iron
- California WET arsenic
- TCLP arsenic

Liquid portions were filtered through a 1.2-μm filter prior to analysis.

Only the solid sedimentation basin residuals samples were used for TCLP and Ca WET arsenic analyses. The results from the chemical characterization analyses are summarized in Table 3.2.

Table 3.2

Settled residuals characterization data from participating utilities

Utility name	Sample description	Solids conc. (percent)	pH	Total As*	Total Fe*	Total Al*	TCLP As (mg/L)	WET As (mg/L)
Louisiana Water Company	Untreated residuals†	4.1	11.5	—	—	—	—	—
(New Iberia), LA	Liquid portion	—	11.3	BDL	460	610	—	—
	Solid portion	100.0	---	110	2,320	11,800	0.012	0.096
City of Great Falls, MT	Untreated residuals†	3.0	7.5	—	—	—	—	—
	Liquid portion	—	7.3	280	290	430	—	—
	Solid portion	100.0	—	490	14,660	109,160	0.158	4.830
City of Helena, MT	Untreated residuals†	3.6	6.7	—	—	—	—	—
	Liquid portion	—	7.7	31	3,720	210	—	—
	Solid portion	100.0	---	420	43,730	41,000	0.125	1.880
Dept. Of Public Utilities	Untreated residuals†	4.1	6.9	—	—	—	—	—
(Billings), MT	Liquid portion	---	7.6	2	1,570	440	—	—
	Solid portion	100.0	---	70	74,670	25,520	0.013	0.863
Lockwood Water Users	Untreated residuals†	4.7	7.2	—	—	—	—	—
Association (Lockwood), MT	Liquid portion	—	7.4	BDL	26,660	400	—	—
	Solid portion	100.0	---	820	327,700	11,800	0.011	0.224
Los Angeles Dept. Of Water	Untreated residuals†	14.0	7.2	—	---	—	—	—
& Power (LADWP), CA	Liquid portion	—	6.8	358	22,010	310	—	—
	Solid portion	100.0	---	730	139,390	16,990	0.162	4.931
Indiana-American Water	Untreated residuals†	1.9	7.2	—	—	—	—	—
Company, IN	Liquid portion	—	7.7	14	500	210	—	—
	Solid portion	100.0	---	15,730	493,200	1,720	0.031	2.471
Heath, OH	Untreated residuals†	26	10	—	—	—	—	—
	Liquid portion	—	9.4	BDL	BDL	2.0	—	—
	Solid portion	100.0	---	19	3,300	9,107	BDL	0.697

BDL= Below Detection Level (< 2 µg/L for As)
*Units for mixed and liquid samples are µg/L; units for solid samples are mg/kg
†Completely mixed residuals before separation

The arsenic concentrations of the secondary liquid and solid residuals are shown graphically in Figures 3.1 and 3.2. The two highest arsenic concentrations in the liquid decant after gravity settling were from LADWP and Great Falls, which were 358 µg/L and 280 µg/L, respectively. The solids arsenic concentrations (Figure 3.2) had a range of between 19 and 15,730 mg/kg.

The toxicity test data are presented in Figure 3.3. These data indicate significant differences between the TCLP arsenic and Ca WET arsenic concentrations measured from the solid residuals samples. For all of the samples tested, the Ca WET arsenic concentration was significantly higher than the TCLP arsenic concentration. These data indicated (using dry solids) that the Ca WET leached arsenic concentration was 8 to 80 times higher compared to the TCLP arsenic leachate. This increase is due to the use of citric acid buffer and anaerobic test conditions for the Ca WET extraction, which is more aggressive for arsenic leaching than acetic acid extraction used for the TCLP. The anaerobic conditions are generated in the Ca WET test by purging the sample with nitrogen gas prior to agitation. These data are consistent with findings from a demonstration study in which an iron coagulant sludge was analyzed using both TCLP and Ca WET extraction procedures to determine arsenic leaching (Amy and Edwards1999). It should be noted that each of the sludges tested comply with the regulatory limit of 5 mg/L using either test. Great Falls and LADWP were, however, only slightly below the maximum limit if the Ca WET test is used for regulatory compliance. All of the TCLP arsenic concentrations were less than 0.2 mg/L.

In the TCLP procedure (Standard Method 1311), for solid residuals samples, one of two extraction fluids is used based on sample pH. The solid media sample plus a volume of extraction fluid equal to 20 times the weight of the sample is added to an extractor vessel, secured in a rotary agitation device and rotated at 20± 2 rpm for 18± 2 hrs. The extract is acidified to less than pH 2 and an acid digestion is performed.

In the Ca WET test extraction (CA SOP 910), 50 grams of sample is placed in the extraction vessel along with 500 ml of the sodium citrate extraction solution. Nitrogen gas is flushed through air scrubbers placed in the mixture for 15 minutes to remove oxygen from the extraction medium. During the 48-hr extraction period, the sample is agitated vigorously. Extracts are filtered and acidified before analysis.

Table 3.3

Key differences in TCLP and Ca WET

Parameter	TCLP	Ca WET
Extraction fluid	Acetic acid	Citric acid
Extraction period	18 hours	48 hours
Anaerobic conditions	No	Yes, by purging with N_2 gas prior to agitation

Some of the key differences in the TCLP and Ca WET extraction procedures are shown below in Table 3.3. Because of these differences, the Ca WET test is a more aggressive leaching test for arsenic.

Additional TCLP and Ca WET tests were performed to evaluate the impact of residuals percent solids concentrations on TCLP and Ca WET arsenic concentrations leached. Two utilities, Great Falls and LADWP (one alum and one ferric residual), were selected for this assessment using residuals collected from the lagoon simulation study after six months of aging. Dewatering and heat drying were used to condition the residuals to desired solids concentrations. The percent solids concentration for each residual evaluated was as follows:

- Great Falls (14-, 25-, 32- and 100-percent solids)
- LADWP (19-, 32-, 41- and 100-percent concentration)

The findings from these tests are provided in Figures 3.4 and 3.5. Figure 3.4 shows that the TCLP arsenic concentrations leached were all less than 0.1 mg/L, with one exception (Great Falls at 14-percent solids concentration). The Great Falls TCLP arsenic concentration decreased with increased solids concentration. The LADWP data (Figure 3.4) indicated that TCLP arsenic leaching increased slightly with increasing solids concentration. Thus, it seems that (a) arsenic leached during both the TCLP and the Ca WET procedures vary with residual percent solids concentration and (b) the relationship between solids concentration and arsenic leaching during these tests is dependent on the quality and nature of those residuals. This relationship has never been explored in the literature as far as the authors are aware, and more research is required.

The Ca WET data (Figure 3.5) for both Great Falls and LADWP demonstrated an increasing trend for arsenic release up to 40-percent solids concentration, followed by a decrease from the 40 percent to the 100-percent solids concentration samples. The two residuals provided very similar trends. In lagoon studies, as described in Chapter 4, arsenic release was greater from 20-percent solids concentration samples compared to dried (100 percent) samples. It is possible that during the drying process a conversion from the hydroxide form of iron to the oxide form, which is more stable, takes place.

SFBW RESIDUALS CHARACTERIZATION

SFBW samples from six different utilities (as shown in Table 3.1) were obtained for characterization and testing. Each SFBW sample was analyzed for total arsenic, total iron, total aluminum, TSS, pH, and turbidity prior to bench-scale testing. Results from these tests are presented in Table 3.4. Analytical methodology is discussed in Chapter 4.

Table 3.4

Results of SFBW characterization analysis

Utility name	Type	Turbidity (ntu)	pH	Total As (µg/L)	Total Fe (mg/L)	Total Al (mg/L)	Ca (mg/L)	Mg (mg/L)	TSS (mg/L)
Louisiana Water Co. (New Iberia), LA	Lime	1,150	8.6	43	18.6	1.94	267	18	2,364
City of Great Falls, MT	Alum	7.8	7.5	32	0.14	4.81	---	---	269
City of Helena, MT	Alum	265	7.1	163	9.2	21.90	---	---	515
Billings, MT	Ferric	28	7.7	63	34.6	1.16	---	---	119
Lockwood Water Users Assoc. (Lockwood), MT	Ferric	12	7.3	14	8.7	0.35	---	---	28
Los Angeles Dept. of Water & Power (LADWP), CA	Ferric	208	7.2	185	66.3	5.63	---	---	

The arsenic concentrations measured in these SFBW samples are shown graphically in Figure 3.6. It should be noted that the Billings, Helena, and LADWP untreated SFBW arsenic concentrations exceeded 50 µg/L, which would most likely prohibit server or water course discharges without pretreatment. This would, however, depend on the limits outlined in each utility's NPDES permit and local POTW limits for these utilities.

WTP MASS BALANCE

In addition to the laboratory characterization work performed on the residuals samples collected, two arsenic mass balance analyses were conducted during the project. The utilities evaluated included Billings Department of Public Utilities (ferric) and Lockwood Water Users Associations WTP (ferric). Both of these plants are located in Montana and use the Yellowstone River as their raw water source. Sampling was conducted by the project team for both of these arsenic mass balance evaluations.

A mass balance schematic for each of these plants is included as Figures 3.7 and 3.8. The arsenic concentration data are shown on the schematic at the locations where only liquid samples were collected, and the arsenic:iron ratios are shown where residuals samples were collected. The contribution of arsenic from the treatment chemicals used is also noted on each schematic and was minor. The arsenic concentrations measured in the ferric chloride, cationic polymer, and nonionic polymer used at the Lockwood WTP were 0.08, 0.12, and 0.01 µg/L, respectively.

During the on-site evaluations, it was determined that an accurate mass balance for either of these WTPs would be extremely difficult to achieve. Many operating parameters vary, and determining the amount of arsenic that is deposited in the lagoons over a specific time period would be impossible. With frequent influent and effluent sampling, accurate coagulation dosing rates, and accurate flow measurements, the total mass of arsenic and iron removed could be determined with a relatively high degree of precision. However, given that the lagoons are unlined and that there are multiple lagoons at each site, it would not be possible to determine the ultimate fate of all of the removed arsenic. It was possible, however, to measure the specific arsenic:iron ratio in the different residuals. The range of the arsenic:iron ratio based on arsenic removal during coagulation ranged from 0.2 to 25 µg/L As per mg/L Fe. Samples from the lagoons collected during this test ranged from below detection levels to 1.5 μg/L As per mg/L Fe. The residuals solids collected in a dry area of the lagoon had similar arsenic concentrations as the samples collected from areas of the lagoon covered with liquid decant.

The Lockwood WTP mass balance data demonstrated that approximately 74 percent of the arsenic loading contributed by the source water was removed by treatment and incorporated in the residuals solids. The source water arsenic loading to the WTP was 34,600 mg As/MG treated, while the finished water arsenic concentration was 2.44 μg/L (9,235 mg As/MG finished water or 26 percent of the raw water loading). The ferric chloride dose applied was 20 mg/L as $FeCl_3$, which is equivalent to an iron feed rate of 26,000 kg Fe^{3+}/MG raw water treated. Based on these measured data, the resulting theoretical arsenic:iron ratio was approximately 1.0 mg As/kg Fe, which matches the results of the solids sample very closely.

The Billings WTP uses the same raw water source as Lockwood, however, the influent arsenic concentration was slightly lower at 8.79 μg/L (33,270 mg As/MG raw water). The ferric chloride dose at Billings was also 20 mg/L as $FeCl_3$ at the time of sampling. The measured influent and effluent arsenic data indicate that approximately 92.6 percent of the arsenic was removed by treatment. The finished water arsenic load was 2,460 mg As/MG. The arsenic removed is incorporated in the sludge solids delivered to the drying beds. The theoretical arsenic:iron ratio for Billings was approximately 1.2 mg As/kg Fe, and this closely matches the measured results.

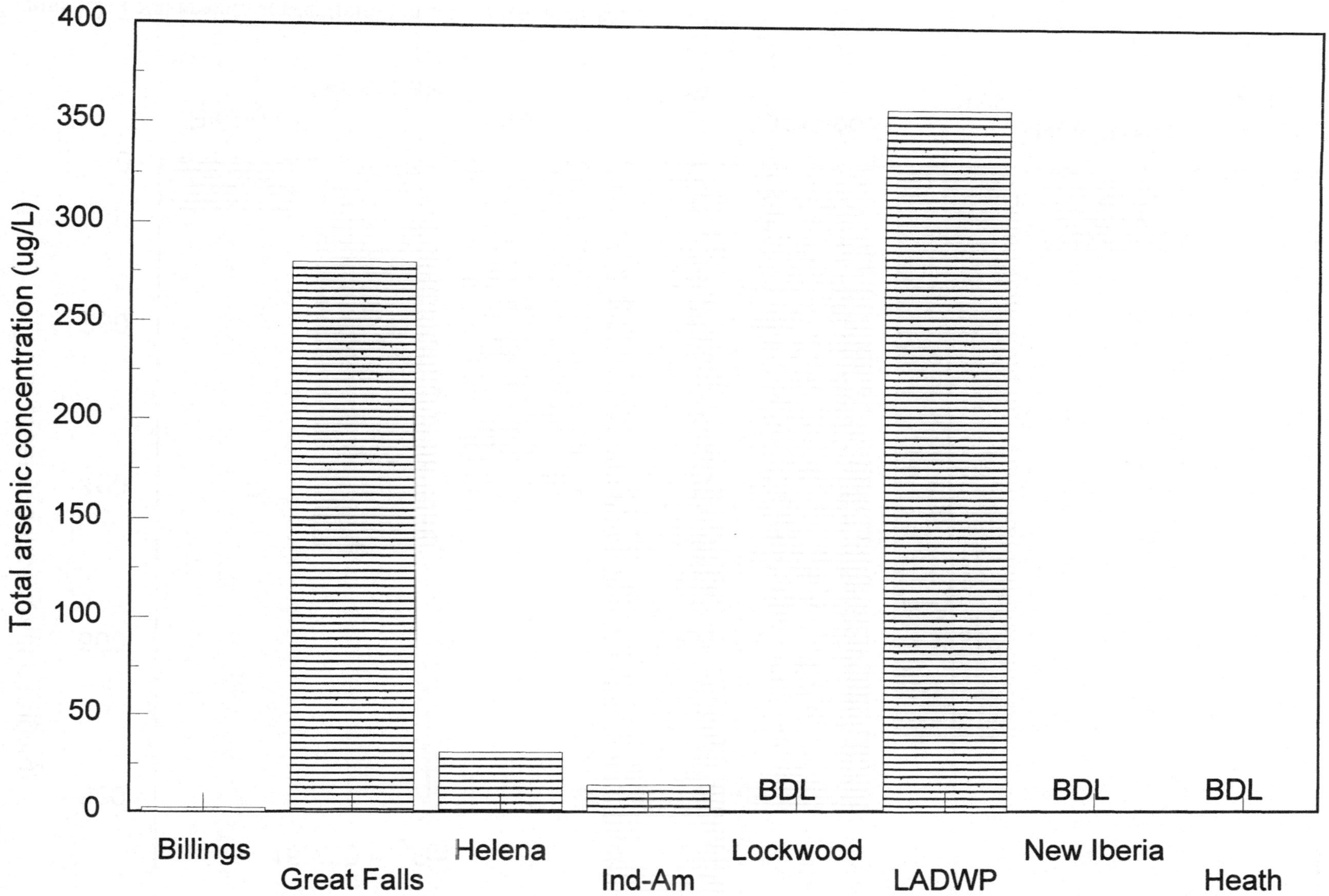

Figure 3.1 Total arsenic concentration in liquid portion of settled residuals after partitioning

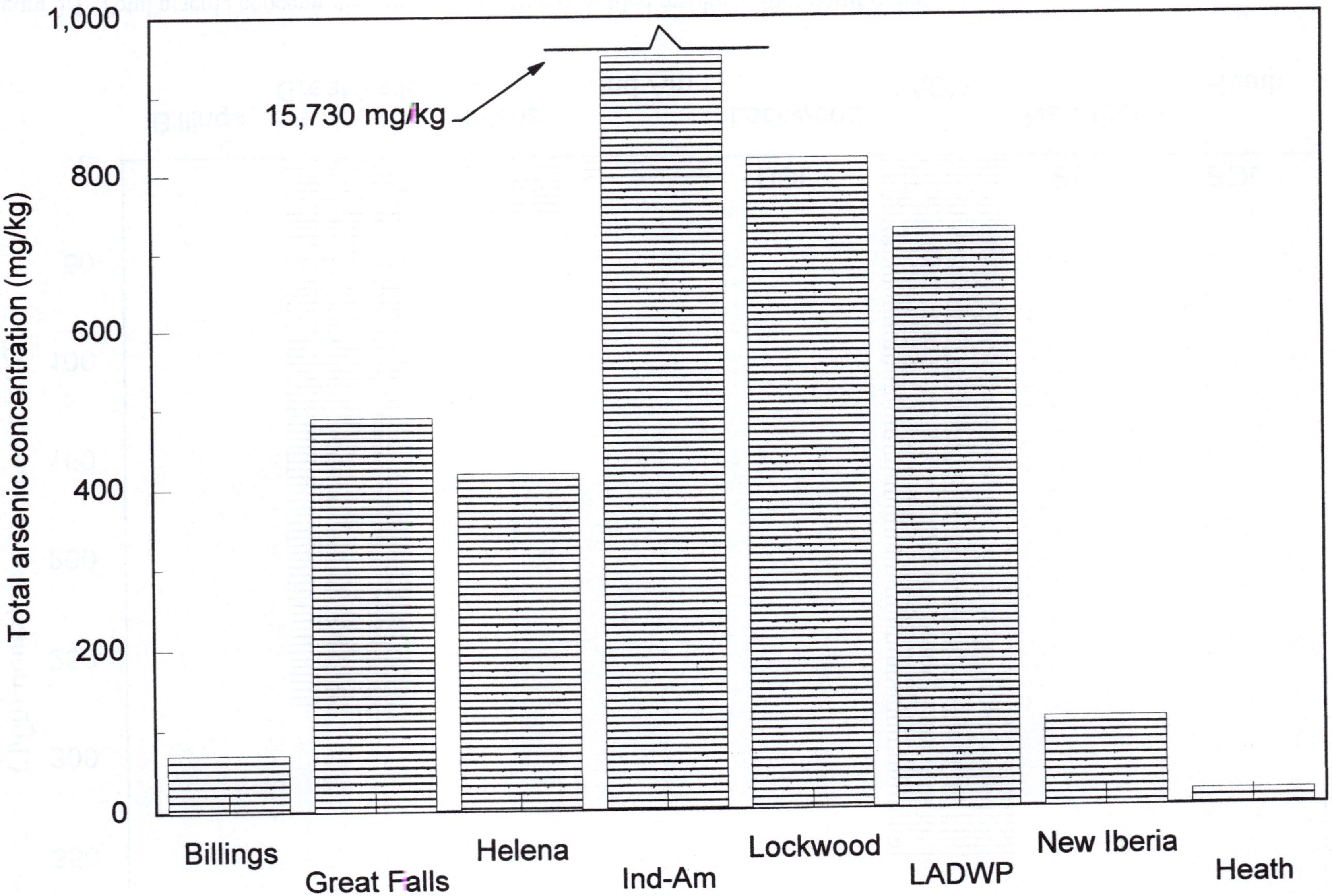

Figure 3.2 Total arsenic concentration in settled residuals solids

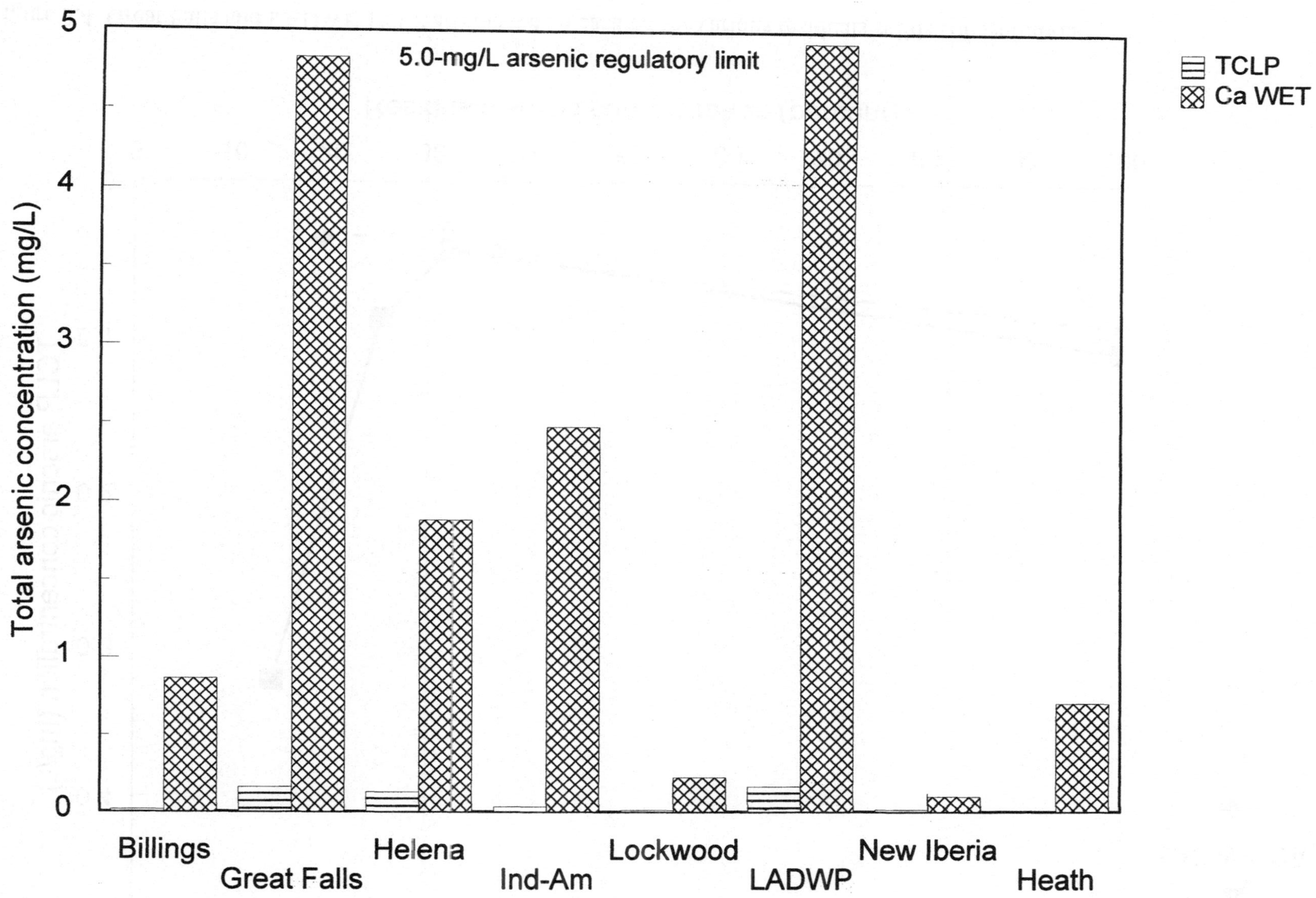

Figure 3.3 TCLP and Ca WET arsenic concentrations for WTP settled residuals solids

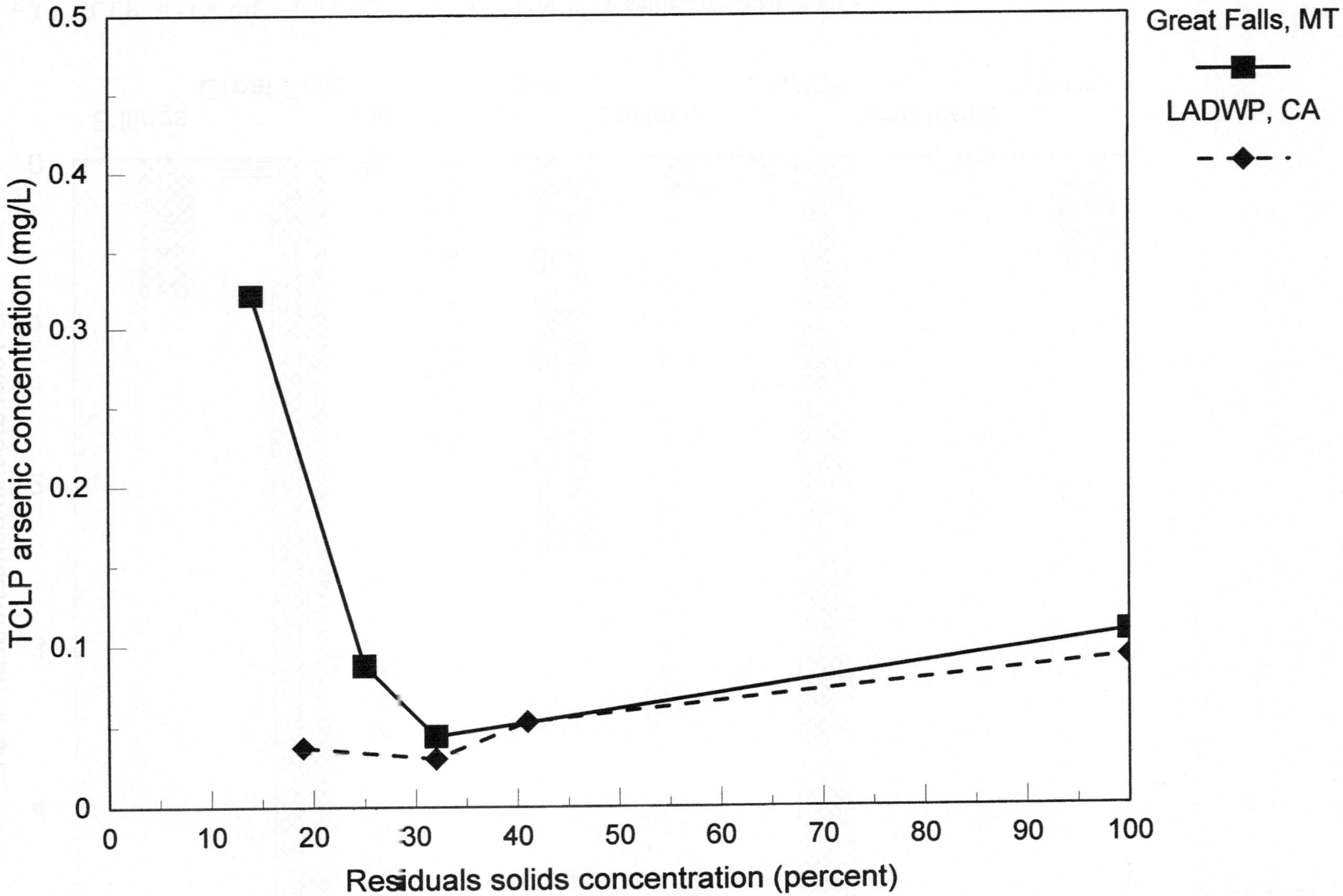

Figure 3.4 Great Falls and LADWP TCLP arsenic data measured for various residuals solids concentrations

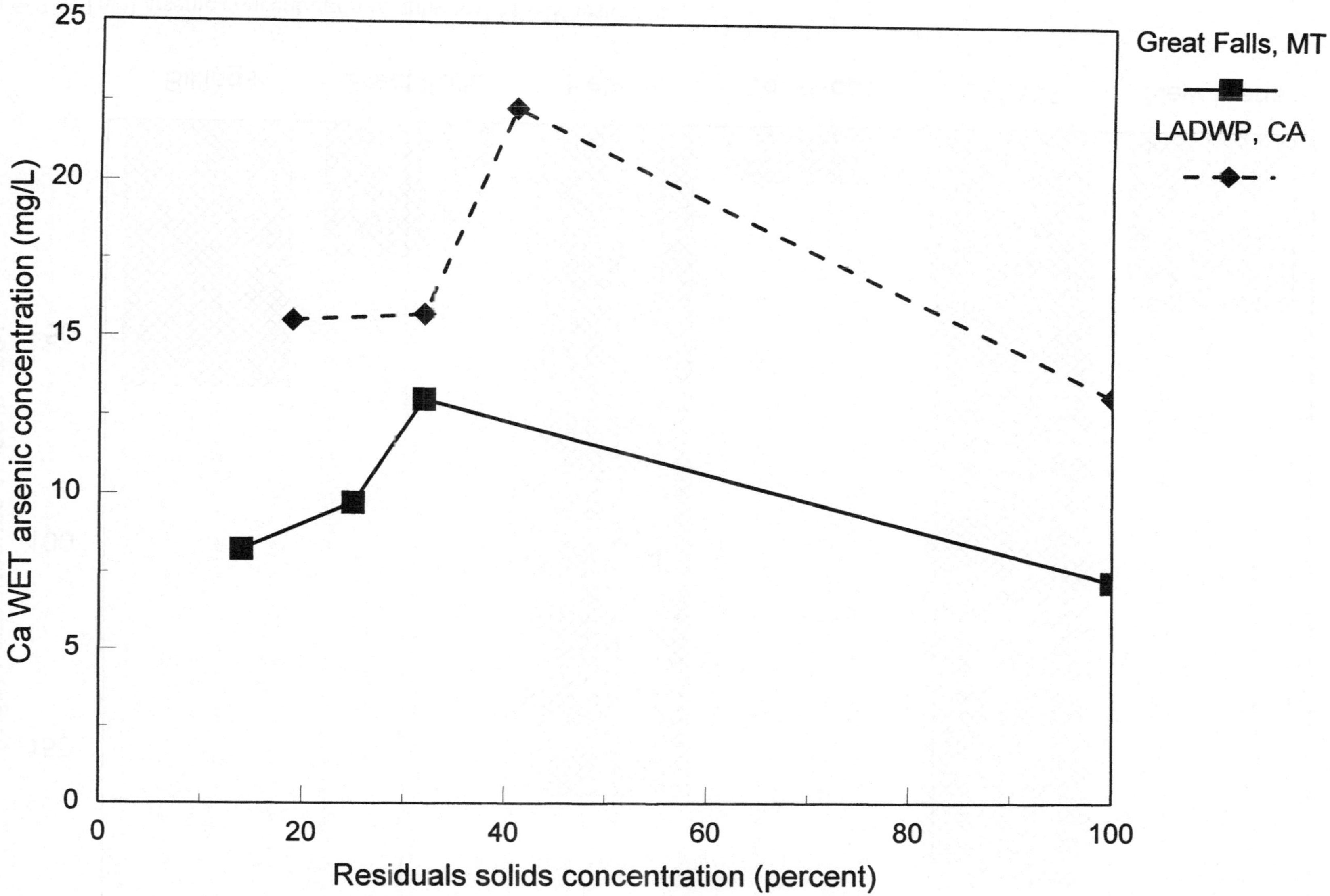

Figure 3.5 Great Falls and LADWP Ca WET arsenic data measured for various residuals solids concentrations

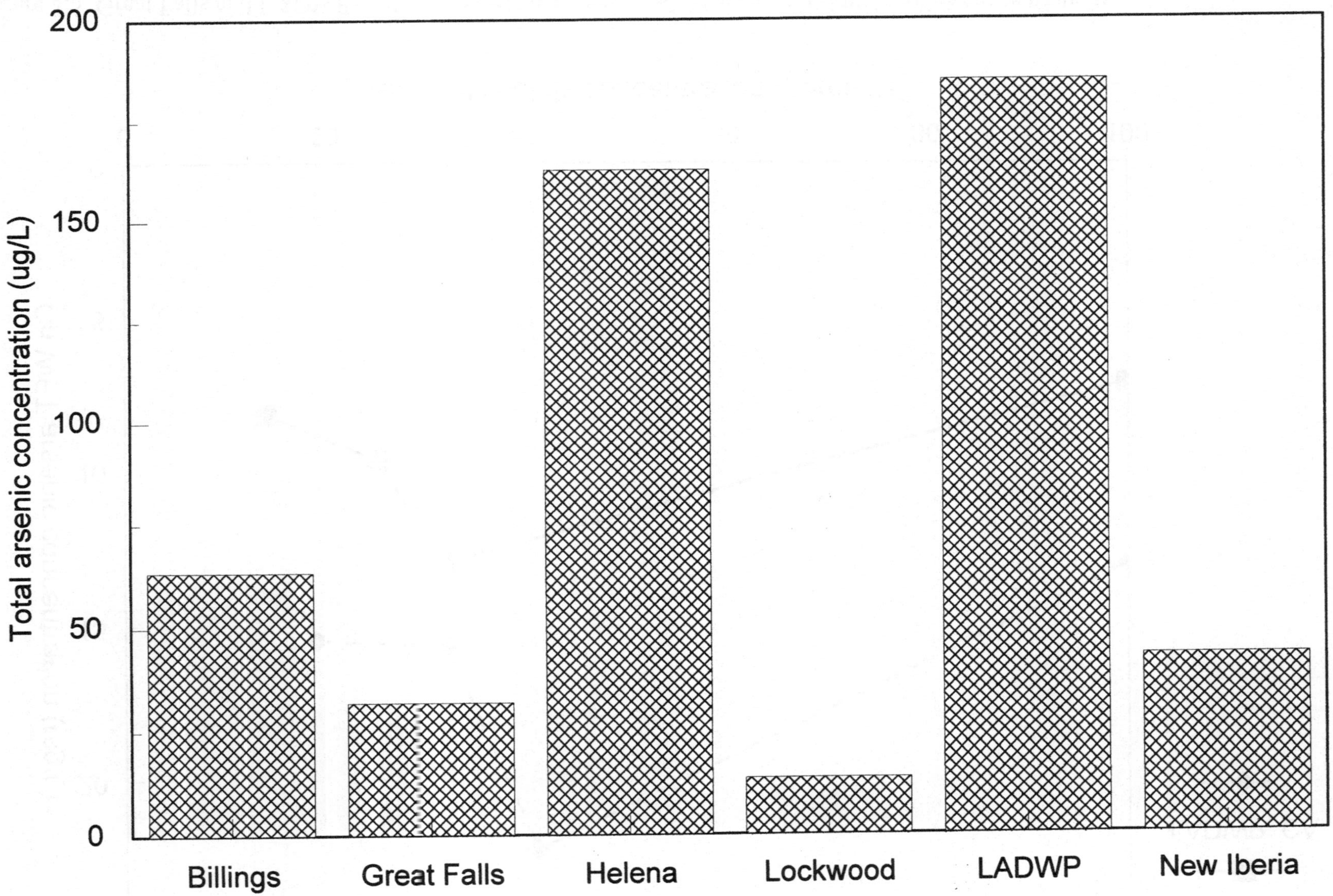

Figure 3.6 Total arsenic concentration in untreated SFBW residuals

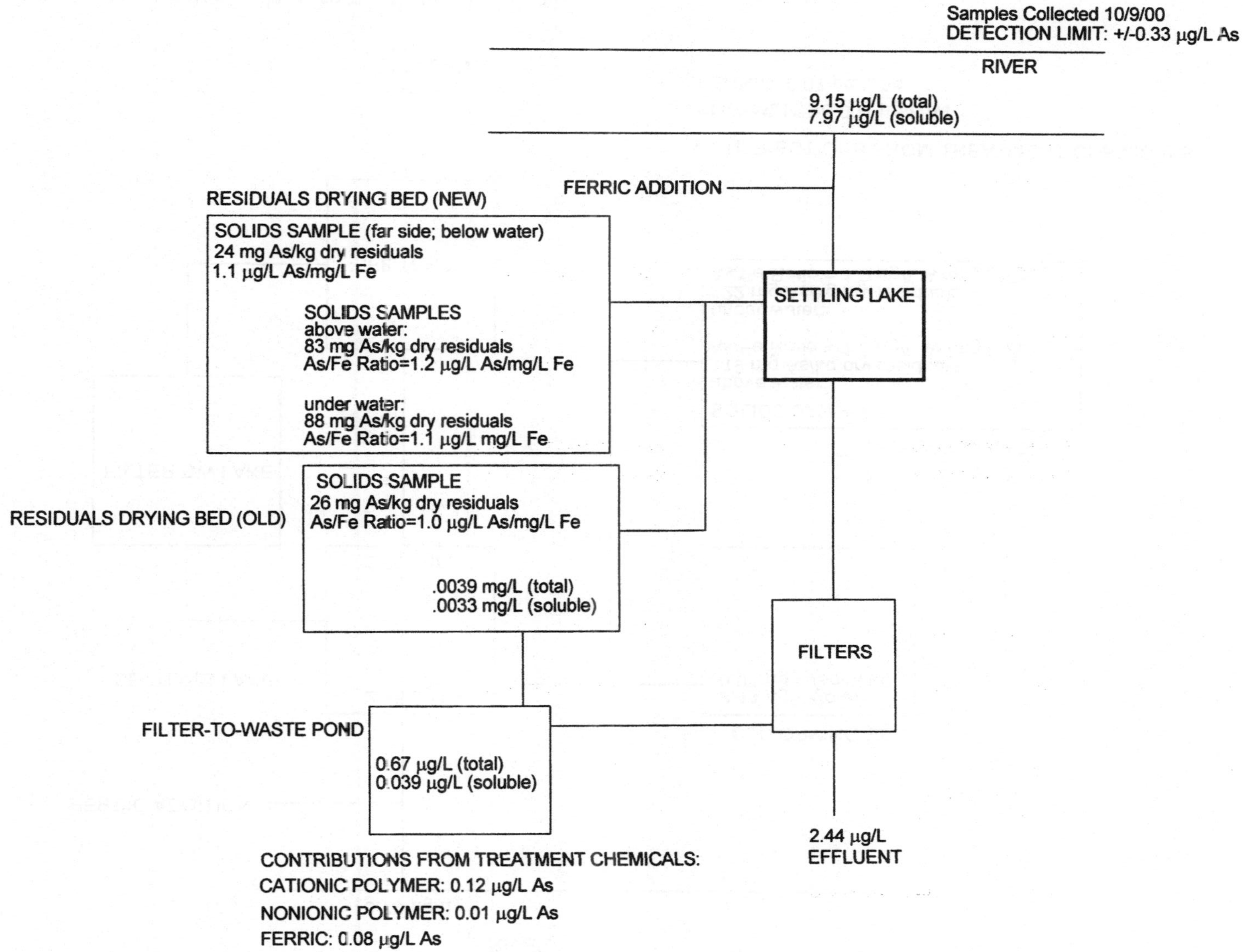

Figure 3.7 Arsenic mass balance data from Lockwood WTP, MT

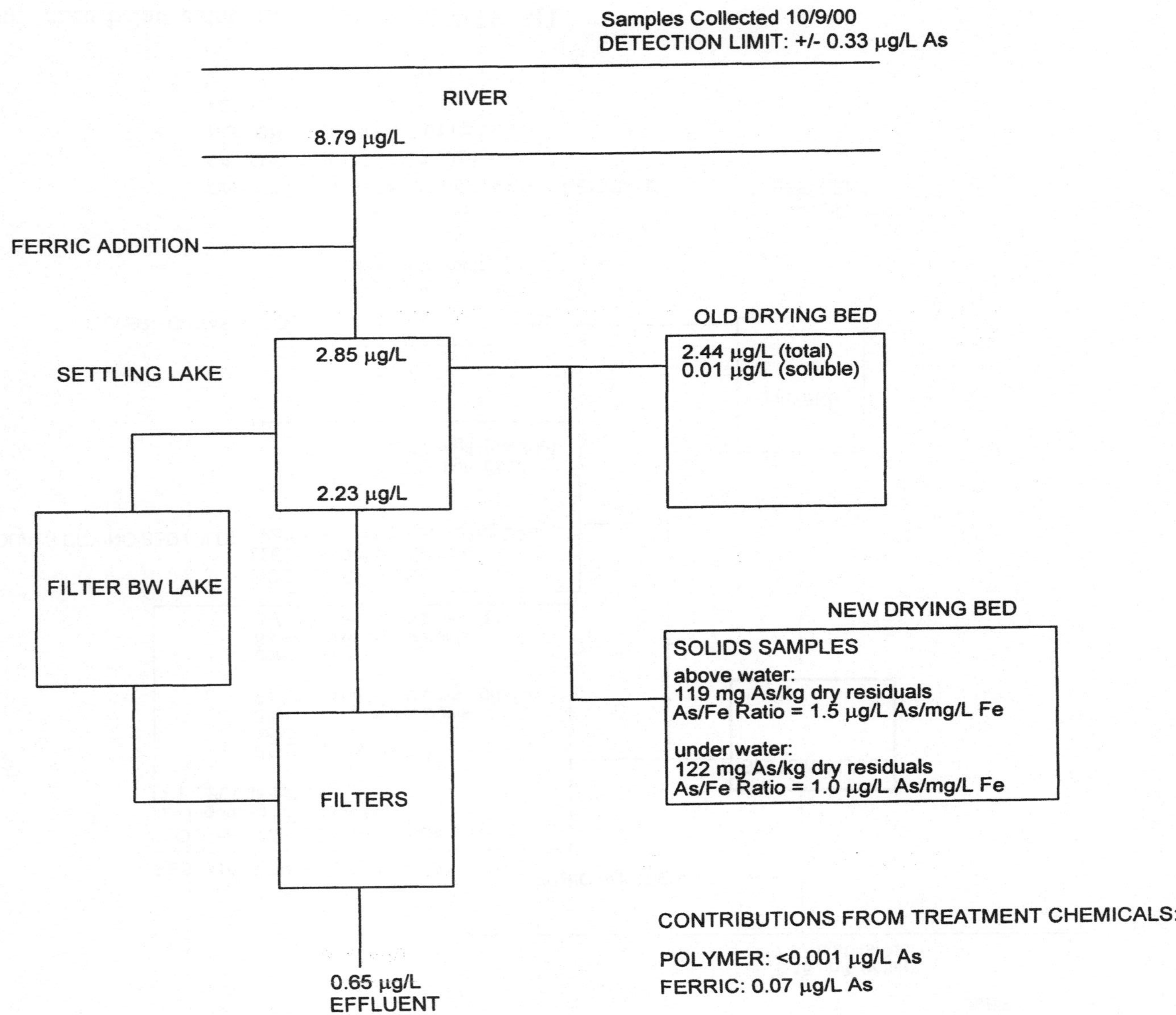

Figure 3.8 Arsenic mass balance data from Billings WTP, MT

CHAPTER 4
LABORATORY TREATABILITY

INTRODUCTION

There is a limited amount of data available that quantifies arsenic concentrations present in drinking water treatment residuals or the potential of arsenic to leach out of residuals during dewatering, lagooning, or after final disposal. This information is critical for selection of residuals handling processes. If the arsenic concentration in residuals leachate has toxic characteristics, disposal options available to a utility would be strictly limited and costly. This study was designed to increase the knowledge base about arsenic mobility in arsenic-containing residuals. The concentrations of arsenic present in each WTP residual stream were measured to establish data for coagulant and softening residuals. These data will help define the disposal options that may be available for As-containing residuals.

The laboratory treatability research tasks conducted during this project included:

- Lagoon simulations
- Sand drying bed simulations
- SFBW clarification using coagulant and polymers
- Lime conditioning and dewatering simulations

Lagoon storage simulations were performed using settled residuals from each of the utilities to determine if and under what conditions arsenic is released. Bench-scale sand drying bed simulations were used to determine the potential of arsenic to leach from settled residuals into the underdrain collection system or possibly into the groundwater. Gravity settling tests, using both ferric chloride and polymer conditioning, were used to remove arsenic from SFBW samples. The clarification techniques used for SFBW treatment are commonly used by WTPs that require solid/liquid separation prior to final disposal. Finally, lime dewatering simulations were conducted to determine the release of arsenic from solid residuals during the dewatering process.

TEST METHODS AND PROCEDURES

Lagoon Simulations

In order to determine how storage time impacts arsenic levels in lagoon leachate or decant, a series of lagoon simulations was developed using nine settled residuals samples. The following three lagoon conditions were simulated:

- Lagoon (no additions)
- Lowered pH lagoon (addition of nitric acid to pH 4.8 using LADWP residuals only)
- Biodegradable organic matter (BOM) lagoon (addition of organics to LADWP residuals only)

It should be noted that all of the samples were used to perform the unaltered lagoon simulation test, while only the LADWP residuals were used for the lowered pH and BOM tests.

Upon delivery of the residuals from each utility, the samples were divided into four 2-liter acid washed glass jars. The first jar of residuals from each utility was tested immediately to obtain the initial contaminant concentrations using "fresh" residuals. The remaining jars were sealed and stored in the dark at 20°C to simulate lagoon storage. The contents of the stored jars were analyzed at two-, four-, and six-month intervals.

A lagoon simulation with added BOM was developed using LADWP residuals. Anaerobic sewage, bactopeptone, glucose, and were added to increase the biological activity and sodium sulfite was added to directly reduce the redox potential as shown in Table 4.1. The LADWP residuals were also acidified using HNO_3 to generate the "reduced pH" lagoon conditions. These lagoon simulations were stored and sampled using the same procedures as the unaltered lagoon simulations at the same time intervals.

Table 4.1

Addition for BOM-lagoon tests

Ingredient	Concentration
Anaerobic sewage	14 ml/L
Bactopeptone	500 mg/L
Glucose	500 mg/L
Sodium sulfite	200 mg/L

The lagoon simulations were initiated in November 2000 for the unaltered and reduced pH test, while the BOM lagoon simulation began in January 2001. During each sampling event the pH, redox potential, and dissolved oxygen (DO) levels were measured, after which the residuals were separated into liquid and solid samples for further laboratory analyses. The residuals were separated into solid and liquid portions using a Whatman #41 filter (20- to 25- μm pore size). The liquid portions were analyzed using Inductively Coupled Plasma-Atomic Emission Spectroscopy (ICP-AES, Methods 3010a and 6010) for total arsenic, total iron, and total aluminum. The solid fractions were also analyzed after drying for 48 hours in a 100°C oven. The dried residual solids were then acid digested (EPA Method 3030) and analyzed for total Fe, Al, and total arsenic using ICP-AES. The arsenic lower detection limit was 2 μg/L.

During the second and sixth month, the solid portions of the filtered residuals were subjected to toxicity testing. The TCLP (EPA Method 1311) and California Waste Extraction Test (Ca WET, Ca HML SOP 910) were performed on each solid residual sample. The TCLP extract was acid digested with EPA Method 3010a. The TCLP digestate and California WET extract were analyzed using an ICP-AES with a hydride generator.

Sand Drying Beds

Pilot-scale sand drying beds were constructed to determine the leaching potential of arsenic from the various sludge samples under different simulated environmental conditions. These test conditions were as follows:

- Typical rain water (pH 5.8)
- Acid rain (pH 3.8 and 4.8)
- Darkness
- Exposure to sunlight

Five sand drying beds were constructed using 6-in. diameter PVC pipe with an attached underdrain for leachate collection. One of the drying beds had no residuals added and served as the control, while the remaining four units each had residuals placed on top of the sand bed (Figure 4.1.)

The settled residuals used in the sand drying bed simulation were from LADWP, which contained the highest concentration of arsenic of the samples received. The four drying beds containing residuals were subject to different environmental and rainfall conditions. Two of the sand drying beds were watered with simulated rainfall at pH 5.8. One of these was exposed to sunlight while the other was covered. The other two sand bed units were watered with simulated acid rain at pH 4.8 and pH 3.8, both exposed to sunlight.

The simulated rain water at pH 5.8 was generated by adding a small amount of sodium chloride to distilled water and adjusting the pH as needed. The solution was purged with nitrogen gas to increase pH, and carbon dioxide was added to decrease pH. The acid rain water at pH 3.8 and 4.8 was created by adding 1 N nitric acid until the desired pH was achieved.

Spent Filter Backwash Clarification

A Phipps and Bird six-paddle stirrer with 2-L Gator jars was used for conducting the spent filter backwash (SFBW) settling tests. The same jar test mixing velocity gradients (G, sec^{-1}) and durations were used for each of the tests. Settling tests were performed on each SFBW using at least four different polymer doses ranging from 0.5 to 5.0 mg/L. Polymer doses were adjusted as necessary to maximize turbidity removal. A 30-sec rapid mix period at a velocity gradient of 550 sec^{-1} was used to disperse the polymer into the SFBW. This mixing condition was determined based on a preliminary mixing intensity investigation conducted using the New Iberia SFBW. Results from these tests (Figure 4.2) demonstrated that a 30-sec mixing period provided the best turbidity removal.

After rapid mixing, the SFBW samples were allowed to settle for up to 30 min. Samples were collected after 2, 5, 10, and 30 min (1.2, 0.5, 0.25, and 0.10 gpm/ft^2 or 1 cm/min) for turbidity analysis. Samples were also collected after 10 min of settling (0.25 gpm/ft^2) for analysis of total arsenic, total iron, total aluminum, and particle counts (>2 μm). Additional settling tests were conducted using both polymer addition and pH reduction using the polymer dose that provided the best turbidity removal. These tests were conducted to determine the effect of pH on both turbidity and arsenic removal.

A second series of tests was performed using ferric chloride for coagulation of the SFBW samples. The same test mixing conditions were used for ferric chloride testing. The doses as $FeCl_3$ used were 20, 40, 80, and 120 mg/L. Adjustment of pH during coagulant testing was only performed when the ambient pH dropped below pH 5.0. The sampling times and analyses conducted for these tests were identical to the polymer tests.

Lime Conditioning

General

Lime is sometimes used as a conditioning chemical for dewatering of coagulant residuals. Therefore, lime was added to two coagulant residuals samples to determine its impact on arsenic release. Residuals used for all of the lime conditioning experiments were obtained from LADWP and Great Falls, MT. Each of these treatment plants has relatively high source water arsenic levels. LADWP uses ferric chloride as a coagulant while Great Falls uses alum. Each residual was evaluated for total solids concentration using Standard Method 2540G, and then acid-digested per Standard Method 3030E to determine the total arsenic concentration (APHA et al. 1998).

Lab grade deionized water was added to each residual sample to obtain a final solids concentration of two percent for testing purposes. Samples were prepared in 250 mL polyethylene bottles by adding an appropriate amount of wet residuals and diluting with lab grade water to a final volume of approximately 125 mL. Calcium hydroxide (lime), sodium hydroxide (caustic soda), or sodium carbonate (soda ash) were added to individual bottles in incremental amounts. Although caustic and soda ash are not used for conditioning, these were added to evaluate arsenic release mechanisms. The lime dose applied ranged from 0 to 20 percent (dry weight). Caustic or soda ash were added to other samples in order to reach the desired pH.

The initial pH was determined with a Corning 313 pH meter. Samples were then placed on an orbital shaker and allowed to react for 30 minutes. The final pH was then determined and the sample was filtered through a 0.45-micron pore size nylon disposable filter. The filter cake was divided into two parts for the subsequent TCLP or Ca WET analysis.

The TCLP test was performed using Standard Method 1311, however, the sample weight was less than 25 grams. Due to this reduced sample weight, a correspondingly lower volume of extraction fluid was used to maintain the required weight ratio. Similarly, the Ca WET was performed per the California Department of Toxic Substances Control Hazardous Material Laboratory SOP 910 guidelines, also using a reduced amount of residual sample.

Calcium Arsenate Solid Formation Test

A 500-mL solution of both calcium chloride and sodium arsenate was prepared at a targeted concentration of 825 mg/L calcium and 2,000 mg/L As. These values were selected because they were the highest concentrations observed during the chemical additive tests. The solution was divided into six (6) aliquots and the pH of each was adjusted using 1-M NaOH to a pH of 8, 9, 10, 11, 12, and 12.5. Each aliquot was placed on an orbital shaker and allowed to react for 30 minutes. After shaking, a portion of each sample was then filtered through a 0.45-micron filter while the remainder was passed through a 0.025-micron filter.

Divalent Cation Theory Test

A two-percent solids concentration by weight solution was prepared using the LADWP residuals. The solution pH was adjusted to 10.0 +/- 0.2 using 1-M sodium hydroxide. Each aliquot was dosed with varying amounts of calcium or magnesium, placed on an orbital shaker for 30 minutes, and then filtered through a 0.45-micron filter. This test was then repeated using the same procedure at pH 11.

Analysis

All samples were analyzed for total arsenic, total iron, total aluminum, Ca, and Mg using an ICP-ES instrument according to Standard Method 3120B with continuous hydride generation (APHA et al. 1998).

TEST RESULTS

Lagoon Simulations

General

The total arsenic, aluminum, and iron concentrations for each residual, the type of coagulant used, and the arsenic to coagulant molar ratio are listed in Table 4.2.

Table 4.2

Residuals metal concentrations and molar ratios

Utility name	Percent solids	As (mg/kg)	Fe (g/kg)	Al (g/kg)	Coagulant	As:Fe molar ratio	As:Al molar ratio
New Iberia, LA	4.1	109	2.3	28.3	Lime	-	-
Great Falls, MT	3.0	490	14.7	109.2	Alum	-	0.002
Helena, MT	3.6	422	43.7	41.0	Alum	-	0.004
Billings, MT	18.6	72	74.7	25.5	FeCl3	0.001	-
Lockwood, MT	4.7	818	327.7	11.8	FeCl3	0.002	-
LADWP, CA	16.1	726	139.4	17.0	FeCl3	0.004	-
Indiana-American, IN	1.9	15,729	4,932.2	1.7	CeCl3	0.002	-

All of the residuals contain elevated arsenic concentrations, most of which would exceed the EPA ceiling limit for land application of biosolids. Therefore, if any of these states follow the 40CFR 503 guidelines for land application of WTP residuals, then the samples could not be land applied.

The entire data set collected from the lagoon simulation study is provided in Table 4.3. Data and findings are discussed in the following sections.

Lagoon pH, DO, and Redox Potential

During the six-month lagoon simulation pH, DO, and redox potential were measured initially and at months 2, 4, and 6. The lagoon simulation jars housing the residuals were closed and airtight, which was intended to generate the anaerobic conditions as would be encountered in the bottom levels of a lagoon.

The lagoon supernatant pH measured over time during the study is provided in Figure 4.3. These data indicate that pH remained fairly constant through the six-month simulation for Helena, Billings, and Lockwood. The Indiana-American, Great Falls, and LADWP supernatant pH levels, however, demonstrated a gradual decrease over time possibly due to biological activity. The Great Falls pH level decreased from 7.5 to 6.8 during the simulation.

The lagoon supernatant DO levels (Figure 4.4) decreased over time due to the microbial activity in the airtight lagoon jars. The DO concentrations measured after four months were less than 1.0 mg/L for all of the residuals, with the exception of the New Iberia lime softening residuals. The DO for the softening residuals was measured at between 7.5 and 8.5 mg/L.

The final water quality parameter measured was the redox potential for the lagoon supernatant (Figure 4.5.) A positive redox potential indicates oxidizing conditions, while a negative redox potential indicates reducing conditions. Reducing conditions are very commonly measured in the bottom of residuals storage lagoons that are subject to anaerobic conditions. The measured data demonstrate that all of the lagoons, with the exception of the Indiana-American (ferric) and New Iberia (softening), had negative redox potentials, after a short period of time.

Table 4.3

Lagoon simulation data by month

Utility name	Sample treatment	Sample age (months)	Sample description	pH	DO (mg/L)	Redox potential (mV)	Total As	Total Fe	Total Al	Ca	Mg	TCLP* As (mg/L)	Ca* WET (mg/L)
							(mg/L combined & liquid: mg/kg solids)						
Dept. of Public Utilities (Billings), MT	None	Fresh	Combined Sludge	6.9	3.2	53	—	—	—	—	—	—	—
			Liquid portion-dissolved metals	—	—	—	BDL	0.03	0.30	—	—	—	—
			Liquid portion-total metals	—	—	—	0.002	1.57	0.44	—	—	—	—
			Solid portion	—	—	—	70	74,670	25,520	—	—	0.013	0.863
		2	Combined Sludge	7.2	1.3	-167	—	—	—	—	—	—	—
			Liquid portion-total metals	—	—	—	0.503	54.72	0.92	—	—	—	—
			Solid portion	—	—	—	—	—	—	—	—	0.010	2.260
		4	Combined Sludge	7.1	0.7	-176	—	—	—	—	—	—	—
			Liquid portion-total metals	---	---	---	0.635	55.76	0.88	—	—	—	—
		6	Combined Sludge	7.1	0.6	-195	—	—	—	—	—	—	—
			Liquid portion-total metals	—	—	—	0.255	13.60	0.68	—	—	—	—
			Solid portion	—	—	—	—	—	—	—	—	0.009	3.187
City of GreatFalls, MT	None	Fresh	Combined Sludge	7.5	2.2	52	—	—	—	—	—	—	—
			Liquid portion-dissolved metals	—	—	—	0.280	BDL	0.43	—	—	—	—
			Liquid portion-total metals	—	—	—	0.280	0.29	0.43	—	—	—	—
			Solid portion	—	—	—	490	14,660	109,160	—	—	0.158	4.830

(continued)

Table 4.3 (continued)

Utility name	Sample treatment	Sample age (months)	Sample description	pH	DO (mg/L)	Redox potential (mV)	Total As	Total Fe	Total Al	Ca	Mg	TCLP* As (mg/L)	Ca* WET (mg/L)
							(mg/L combined & liquid: mg/kg solids)						
		2	Combined Sludge	6.9	0.6	-112	—	—	—	—	—	—	—
			Liquid portion-total metals	—	—	—	0.547	0.21	1.44	—	—	—	—
			Solid portion	—	—	—	—	—	—	—	—	0.048	6.110
		4	Combined Sludge	7.0	0.9	-133	—	—	—	—	—	—	—
			Liquid portion-total metals	—	—	—	0.354	0.22	1.24	—	—	—	—
		6	Combined Sludge	6.8	0.6	-145	—	—	—	—	—	—	—
			Liquid portion-total metals	—	—	—	—	—	—	—	—	—	—
			Solid portion	—	—	—	—	—	—	—	—	0.109	7.183
City of Helena, MT	None	Fresh	Combined Sludge	6.7	1.8	116	—	—	—	—	—	—	—
			Liquid portion-dissolved metals	—	—	—	0.008	0.31	0.17	—	—	—	—
			Liquid portion-total metals	—	—	—	0.031	3.72	0.21	—	—	—	—
			Solid portion	—	—	—	420	43,730	41,000	—	—	0.125	1.880
		2	Combined Sludge	7.1	0.4	-61	—	—	—	—	—	—	—
			Liquid portion-total metals	—	—	—	0.061	18.97	3.48	—	—	—	—
			Solid portion	—	—	—	—	—	—	—	—	0.130	5.610
		4	Combined Sludge	6.9	0.5	-95	—	—	—	—	—	—	—
			Liquid portion-total metals	—	—	—	0.066	40.69	0.59	—	—	—	—

(continued)

Table 4.3 (continued)

Utility name	Sample treatment	Sample age (months)	Sample description	pH	DO (mg/L)	Redox potential (mV)	Total As	Total Fe	Total Al	Ca	Mg	TCLP* As (mg/L)	Ca* WET (mg/L)
							(mg/L combined & liquid: mg/kg solids)						
		6	Combined Sludge	6.9	0.5	-149	—	—	—	—	—	—	—
			Liquid portion-total metals	—	—	—	0.078	43.76	0.55	—	—	—	—
			Solid portion	—	—	—	—	—	—	—	—	0.144	10.296
Indiana-American Water Co., IN	None	Fresh	Combined Sludge	7.2	4.0	287	—	—	—	—	—	—	—
			Liquid portion-dissolved metals	—	—	—	0.003	0.06	0.17	—	—	—	—
			Liquid portion-total metals	—	—	—	0.014	0.50	0.21	—	—	—	—
			Solid portion	—	—	—	15,730	647,200	1,720	—	—	0.031	2.471
		2	Combined Sludge	7.1	2.3	194	—	—	—	—	—	—	—
			Liquid portion-total metals	—	—	—	0.102	2.20	0.99	—	—	—	—
			Solid portion	—	—	—	—	—	—	—	—	0.010	10.230
		4	Combined Sludge	7.0	1.0	214	—	—	—	—	—	—	—
			Liquid portion-total metals	—	—	—	0.100	2.52	0.45	—	—	—	—
		6	Combined Sludge	6.9	0.7	187	—	—	—	—	—	—	—
			Liquid portion-total metals	—	—	—	0.152	4.93	0.45	—	—	—	—
			Solid portion	—	—	—	—	—	—	—	—	0.020	5.380
Lockwood Water Users Assoc. (Lockwood), MT	None	Fresh	Combined Sludge	7.2	1.2	-81	—	—	—	—	—	—	—
			Liquid portion-dissolved metals	—	—	—	BDL	0.49	0.31	—	—	—	—

(continued)

Table 4.3 (continued)

Utility name	Sample treatment	Sample age (months)	Sample description	pH	DO (mg/L)	Redox potential (mV)	Total As	Total Fe	Total Al	Ca	Mg	TCLP* As (mg/L)	Ca* WET (mg/L)
							(mg/L combined & liquid: mg/kg solids)						
			Liquid portion-total metals	—	—	—	BDL	26.66	0.40	—	—	—	—
			Solid portion	—	—	—	820	231,640	11,800	—	—	0.011	0.224
		2	Combined Sludge	7.4	0.5	-110	—	—	—	—	—	—	—
			Liquid portion-total metals	—	—	—	0.027	25.90	0.32	—	—	—	—
			Solid portion	—	—	—	—	—	—	—	—	0.009	0.260
		4	Combined Sludge	7.2	0.5	-136	—	—	—	—	—	—	—
			Liquid portion-total metals	—	—	—	0.017	18.92	0.50	—	—	—	—
		6	Combined Sludge	7.2	0.4	-157	—	—	—	—	—	—	—
			Liquid portion-total metals	—	—	—	0.014	22.53	0.32	—	—	—	—
			Solid portion	—	—	—	—	—	—	—	—	0.013	0.266
Los Angeles Dept. of Water & Power (LADWP), CA	None	Fresh	Combined Sludge	7.2	1.2	-116	—	—	—	—	—	—	—
			Liquid portion-dissolved metals	—	—	—	BDL	0.46	0.19	—	—	—	—
			Liquid portion-total metals	—	—	—	0.358	22.01	0.31	—	—	—	—
			Solid portion	—	—	—	730	139,390	16,990	—	—	0.162	4.931
		2	Combined Sludge	7.1	0.4	-150	—	—	—	—	—	—	—
			Liquid portion-total metals	—	—	—	0.870	69.56	0.33	—	—	—	—
			Solid portion	—	—	—	—	—	—	—	—	0.079	15.240

(continued)

Table 4.3 (continued)

Utility name	Sample treatment	Sample age (months)	Sample description	pH	DO (mg/L)	Redox potential (mV)	Total As (mg/L combined & liquid: mg/kg solids)	Total Fe	Total Al	Ca	Mg	TCLP* As (mg/L)	Ca* WET (mg/L)
		4	Combined Sludge	7.1	0.4	-142	—	—	—	—	—	—	—
			Liquid portion-total metals	—	—	—	0.486	57.37	0.40	—	—	—	—
		6	Combined Sludge	6.9	0.5	-171	—	—	—	—	—	—	—
			Liquid portion-total metals	—	—	—	0.327	32.23	0.25	—	—	—	—
			Solid portion	—	—	—	—	—	—	—	—	0.093	13.10
Los Angeles Dept. of Water & Power (LADWP), CA	Lowered pH	Fresh	Combined Sludge	4.8	2.3	11	—	—	—	—	—	—	—
			Liquid portion-dissolved metals	—	—	—	2.721	337.5	1.16	—	—	—	—
			Liquid portion-total metals	—	—	—	2.721	337.5	1.16	—	—	—	—
			Solid portion	—	—	—	1,040	219,950	13,800	—	—	0.032	5.856
		2	Combined Sludge	3.1	2.2	331	—	—	—	—	—	—	—
			Liquid portion-total metals				1.487	3854.0	2.06	—	—	—	—
			Solid portion	—	—	—	—	—	—	—	—	N/A	NA
		4	Combined Sludge	4.4	2.3	63	—	—	—	—	—	—	—
			Liquid portion-total metals				1.193	8354.2	2.66	—	—	—	—
		6	Combined Sludge	4.9	1.7	41	—	—	—	—	—	—	—
			Liquid portion-total metals	—	—	—	1.140	5956.00	2.38	—	—	—	—
			Solid portion	—	—	—	—	—	—	—	—	0.057	4.395

(continued)

Table 4.3 (continued)

Utility name	Sample treatment	Sample age (months)	Sample description	pH	DO (mg/L)	Redox potential (mV)	Total As	Total Fe	Total Al	Ca	Mg	TCLP* As (mg/L)	Ca* WET (mg/L)
							(mg/L combined & liquid: mg/kg solids)						
Los Angeles Dept. of Water & Power (LADWP), CA	BOM	Fresh	Combined Sludge	7.0	1.4	-220	—	—	—	—	—	—	—
			Liquid portion-dissolved metals	—	—	—	0.243	20.66	0.26	—	—	—	—
			Liquid portion-total metals	—	—	—	0.493	30.87	0.32	—	—	—	—
			Solid portion	—	—	—	0.530	155.82	11.86	—	—	0.093	2.365
		2	Combined Sludge	7.0	1.0	-82	—	—	—	—	—	—	—
			Liquid portion-total metals	—	—	—	1.045	39.75	0.35	—	—	—	—
			Solid portion	—	—	—	—	—	—	—	—	N/A	N/A
		4	Combined Sludge	7.0	0.7	-137	—	—	—	—	—	—	—
			Liquid portion-total metals	—	—	—	1.100	23.51	0.28	—	—	—	—
		6	Combined Sludge	N/A	N/A	N/A	—	—	—	—	—	—	—
			Liquid portion-total metals	—	—	—	0.295	1.40	BDL	—	—	—	—
			Solid portion	—	—	—	—	—	—	—	—	0.219	23.457
Louisiana Water Co. (New Iberia), LA	None	Fresh	Combined Sludge	11.5	8.3	33	—	—	—	—	—	—	—
			Liquid portion-dissolved metals	—	—	—	BDL	0.46	0.52	54	0.12	—	—
			Liquid portion-total metals	—	—	—	BDL	0.46	0.61	61	0.44	—	—
			Solid portion	—	—	—	110	2,320	11,800	195,160	28,350	0.012	0.096

(continued)

Table 4.3 (continued)

Utility name	Sample treatment	Sample age (months)	Sample description	pH	DO (mg/L)	Redox potential (mV)	Total As	Total Fe	Total Al	Ca	Mg	TCLP* As (mg/L)	Ca* WET (mg/L)
							(mg/L combined & liquid: mg/kg solids)						
		2	Combined Sludge	11.7	7.2	178	—	—	—	—	—	—	—
			Liquid portion-total metals	—	—	—	0.005	0.03	0.23	62.4	0.29		
			Solid portion	—	—	—	—	—	—	—	—	0.010	0.050
		4	Combined Sludge	11.7	7.2	107	—	—	—	—	—	—	—
			Liquid portion-total metals	—	—	—	0.003	0.04	0.26	83.8	0.19	—	—
		6	Combined Sludge	11.5	8.0	132	—	—	—	—	—	—	—
			Liquid portion-total metals	—	—	—	BDL	BDL	0.19	66.0	0.51	—	—
			Solid portion	—	—	—	—	—	—	—	—	0.01	.108

BDL = Below Detection Level (0.002 mg/L As, 0.03 mg/L aluminum)
N/A = Not analyzed
*Tested at 100-percent solids concentration

Ferric Chloride Residuals

A total of four of the utility samples used for lagoon simulation testing were ferric residuals. During testing, the lagoon supernatant arsenic concentration was measured over time to demonstrate the effect of lagoon conditions on arsenic release from solid to liquid phase. These data are provided in Figure 4.6. The figure demonstrates that arsenic leaching from the lagoon solids occurred soon after lagoon simulation testing was initiated. All residuals, except Lockwood, MT, demonstrated increased levels of arsenic in the supernatant. The Billings lagoon had the highest increase in total arsenic from an initial concentration of 2 μg/L to 500 μg/L after two months of storage. The Billings and LADWP lagoons demonstrated a rapid increase in total arsenic and then a decrease back towards the initial arsenic concentration after six months of storage. The authors hypothesize that under the septic, low redox conditions, release of sulfur from decayed organics and reprecipitation of arsenic as arsenic sulfide occurred. Members of the research team are further exploring this possibility under separate research efforts.

The total arsenic concentration of the lagoon supernatant very closely followed the trend of total iron release from the lagoon solids. The iron data are included in Figure 4.7. The iron in the Billings and LADWP residuals was quickly released into the lagoon supernatant and then returned to initial levels after six months. These data indicate that the arsenic concentrations in solution are related to iron releases into solution.

Figure 4.8 is a plot of the total arsenic and iron concentrations in the ferric chloride residual lagoons. The arsenic and iron concentrations correlate well for all of the ferric chloride residuals evaluated except Lockwood, MT. Disregarding the Lockwood data, the R^2 value for these data is approximately 0.89. Figure 4.9 is plot of the total arsenic concentrations versus the change in total iron concentrations in the lagoon supernatant. The change in total iron is the difference between the iron concentration at time zero and the values after storage. This graph demonstrates that the total arsenic concentrations are related to the change in total iron and not necessarily the total concentration of iron. The R^2 value of this plot is 0.89.

The impacts of reducing conditions on metals release from residuals during lagoon simulation testing are demonstrated in Figures 4.10 and 4.11. Figure 4.10 shows the effects of oxidation-reduction conditions on release of iron from the solid to liquid phase. The figure clearly demonstrates that under reducing conditions iron was released into solution at high concentrations. The Indiana-American lagoon was not subjected to a negative redox potential and as a result had a lower concentration of iron in the lagoon supernatant. The Indiana-American residuals contained appreciable concentrations of both iron and arsenic, so it is likely that there was little organic matter in the waste to create reducing conditions. The total arsenic concentration versus redox potential is shown in Figure 4.11. That figure also demonstrates that elevated arsenic levels generally formed in the lagoon supernatant due to the reducing conditions. The Lockwood lagoon, however, had a negative redox potential of between -75 and -150 mV but did not demonstrate significant increases in arsenic or iron concentrations.

Alum Residuals

The two alum residuals tested during the lagoon simulation study were Great Falls, MT and Helena, MT. The arsenic concentrations in the lagoon supernatant over time are shown in Figure 4.12. The figure indicates that the arsenic concentration slightly increased over time for both residuals. The Great Falls data demonstrated a large increase in arsenic from 300 μg/L to almost 600 μg/L at two months, while the total arsenic for the four-month and six-month sample events were measured at initial levels. The Helena lagoon soluble arsenic concentration only slightly increased over time.

The Helena and Great Falls iron data versus time is shown in Figure 4.13. The Great Falls data exhibited an increasing trend in total iron over time from less than 5 mg/L to almost 45 mg/L, while no iron release was measured from the Helena residuals. Because alum was used as the coagulant for these residuals, the aluminum release was also measured. These data are provided in Figure 4.14. The plot shows that total aluminum increased as a result of lagoon storage conditions. The Great Falls lagoon aluminum release increased over time, while the Helena aluminum concentration spiked after two months to 3.5 mg/L and then returned to around 0.5 mg/L by the month four-sample event. Based on the data in Figures 4.12, 4.13, and 4.14, it appears that even though these residuals were generated by alum coagulation, the reduction and release of iron was associated with arsenic release into solution.

Similar to the ferric residuals analysis, the alum lagoon supernatant arsenic and iron concentrations were plotted versus the lagoon redox potentials. The effect of redox potential on iron release from the Helena and Great Falls residuals is shown in Figure 4.15. These data indicate that each of these two lagoons developed reducing conditions with redox potentials of between -50 and -150 mV. The supernatant iron concentration in the Helena lagoon increased as redox potential decreased, and the arsenic concentration (Figure 4.16) remained constant. The Great Falls lagoon, however, only exhibited a very small release of iron under reducing conditions, but demonstrated an increase in arsenic leaching. The Great Falls residuals only had an initial iron concentration of 15,000 mg/kg compared to 44,000 mg/kg of iron in the Helena residuals. These data again suggest that significant leaching of arsenic could occur in lagoons that experience reducing conditions.

Lime Softening Residuals

The only lime softening residual used for lagoon simulation testing was from New Iberia, La. The lagoon test results (summarized previously in the Table 4.3) demonstrated that there was no release of arsenic to the lagoon supernatant under lagoon storage conditions. The lagoon pH remained at 11.7, and the redox potential after storage increased to as high as 178 mV. Iron and aluminum concentrations in the lagoon supernatant residuals stream were initially low, and no leaching over time was noted for either of these metals. Due to the high buffering capacity of lime softening residuals, arsenic released under lagoon storage conditions is not expected to be a problem.

Modified Lagoon Conditions

The LADWP ferric residuals were used to develop two additional lagoon simulation tests that included a reduced pH lagoon condition and a lagoon with added biodegradable organic matter (BOM). The pH conditions during the six month simulation are provided in Figure 4.17. These data show that the initial pH of the unaltered and BOM lagoons were approximately 7.0, while the reduced pH lagoon was less than five. The pH of the BOM and unaltered lagoon simulations remained constant during the test period, while the reduced pH lagoon decreased to approximately pH 3.9 after two months of storage and then returned to around pH 5.0 during the four- and six-month sample events.

The redox potentials of the LADWP lagoons are shown in Figure 4.18. These data demonstrate that both the unaltered and BOM lagoon had reducing conditions (-100 to -200 mV) while the reduced pH lagoon remained redox positive.

Leaching of iron from the LADWP residuals during the study is shown in Figure 4.19. The data show that the addition of acid to reduce the pH resulted in an immediate release of iron into solution of from 300 mg/L to as high as 8,400 mg/L. The BOM and unaltered lagoons had soluble iron concentrations of less than 100 mg/L.

The lagoon supernatant arsenic data (Figure 4.20) demonstrate that the reduced pH and BOM additions to the LADWP residuals also increased arsenic release. The arsenic concentration in the reduced pH lagoon was immediately almost 2.75 mg/L after acid addition compared to 0.36 mg/L arsenic in the unaltered lagoon. The maximum total arsenic concentrations in the BOM and unaltered lagoons were 1.1 mg/L and 0.87 mg/L, respectively.

Lagoon TCLP and Ca WET Results

Solids samples were collected from the lagoon simulations after two and six months of storage time for TCLP and Ca WET arsenic analyses. These data are shown in Table 4.4 along with the results from the fresh residuals sample analyzed prior to lagoon simulation testing.

Table 4.4

TCLP and Ca WET results from lagoon samples

	TCLP arsenic (mg/L)			Ca WET arsenic (mg/L)		
Utility name	Fresh	Aged 2 months	Aged 6 months	Fresh	Aged 2 months	Aged 6 months
Louisiana Water Co. (New Iberia), LA	0.012	0.010	0.010	0.096	0.05	0.11
City of Great Falls, MT	0.158	0.048	0.109	4.830	6.11	7.18

(continued)

Table 4.4 (continued)

Utility name	TCLP arsenic (mg/L)			Ca WET arsenic (mg/L)		
	Fresh	Aged 2 months	Aged 6 months	Fresh	Aged 2 months	Aged 6 months
City of Helena. MT	0.125	0.130	0.144	1.880	5.61	10.3
Dept. of Public Utilities (Billings), MT	0.013	0.010	0.009	0.863	2.26	3.19
Lockwood Water Users Assoc. (Lockwood), MT	0.011	0.009	0.013	0.224	0.26	0.27
Los Angeles Dept. of Water & Power (LADWP), CA	0.162	0.079	0.093	4.931	15.24	13.1
Indiana-American Water Co., IN	0.031	0.010	0.020	2.470	10.23	5.38

Note: TCLP and Ca WET test results in this table were performed using 100-percent dry solids

Fresh residuals TCLP and Ca WET results are plotted with 6-month data in Figures 4.21 and 4.22. Figure 4.21 shows the TCLP arsenic results in comparison to the 5.0-mg/L regulatory limit for toxicity determination. The results show that each of the lagoon simulation solid residuals would meet the TCLP arsenic limit. The maximum arsenic concentration measured was approximately 0.15 mg/L. The Ca WET arsenic data, however, present a different picture in terms of arsenic leaching. These data (shown in Figure 4.22) indicate that two of the fresh residuals were only slightly below the Ca WET arsenic limit of 5.0 mg/L, and after lagoon storage for six months, four of the six residuals exceeded the 5.0-mg/L Ca WET arsenic limit.

Figure 4.23 illustrates the impact of aging for two and six months on Ca WET and TCLP arsenic levels. The log-scale plot demonstrates graphically how much higher Ca WET concentrations were compared to TCLP levels and also shows that Ca WET arsenic levels generally increased from two to six months. TCLP levels, as shown in the plot and in Table 4.4, were fairly similar at two and six months.

Lagoon Measured Arsenic Versus TCLP and Ca WET Arsenic Release

The method generally used by a water utility to evaluate toxicity of WTP residuals is the TCLP test, unless the utility is in California and the Ca WET test is required. The TCLP test (and Ca WET test in California) is intended to predict release of toxic contaminants from solid wastes when landfilled, land applied, lagooned, or for other environmental applications. One of the goals of this project was to compare TCLP and Ca WET arsenic leaching with measured arsenic concentrations collected from an actual lagoon simulation. This summary attempts to correlate the findings from the toxicity tests with the measured pilot lagoon releases to determine how well these tests predict arsenic release for this type of application.

In order to normalize the lagoon, TCLP, and Ca WET arsenic release data for comparison, a series of calculations had to be performed to account for the differences in lagoon and toxicity test residuals solids concentrations and the extraction volumes used. The total mass of arsenic released per unit weight of dry solids used for the Ca WET and TCLP tests was calculated and compared with the results from the pilot lagoon simulation. These data are plotted in Figures 4.24 and 4.25 to demonstrate the difference between TCLP and Ca Wet arsenic release versus actual measured arsenic release from the lagoon tests.

Figure 4.24 shows the lagoon arsenic release versus the TCLP arsenic release (using both 20-percent and 100-percent solids concentration data from TCLP analysis). These data show that the TCLP (20 percent solids) test resulted in lower arsenic release for three of the residual samples compared to measured lagoon concentrations using the 2-month data and higher release for the remaining four. The ratios of TCLP arsenic release (at both 20 and 100 percent) at two months to lagoon arsenic release are shown in Table 4.5. The data exceeding 1.0 indicate that the TCLP leaching was greater than the lagoon arsenic leaching. The data tables and calculations used to develop these ratio comparisons between the toxicity tests and lagoon simulation are provided in Appendix Tables B.1, B.2, and B.3.

Table 4.5

Ratio of TCLP and Ca WET to arsenic lagoon release at two months

	Ca WET: Lagoon		TCLP: Lagoon	
Utility	20 percent	Dry	20 percent	Dry
Louisiana Water Co. (New Iberia), LA	63	3.7	5.00	15.8
City of Great Falls, MT	20	3.3	2.30	0.05
City of Helena, MT	8	3.1	0.20	0.14
Dept. of Public Utilities (Billings), MT	156	6.6	1.90	0.06
Lockwood Water Users Assoc. (Lockwood), MT	362	4.6	1.80	0.35
Los Angeles Dept. of Water & Power (LADWP), CA	8	6.9	0.04	0.01
Indiana-American Water Co., IN	694	19.0	0.30	0.04

The Ca WET data (Figure 4.25) demonstrate a much higher arsenic release than for the lagoon simulations. For all of the residuals tested, the Ca WET test yielded higher arsenic concentrations compared with actual lagoon simulation arsenic releases. The ratios of Ca WET arsenic to measured lagoon arsenic are also provided in Table 4.5. The Ca WET test (at 20- percent solids concentration) resulted in arsenic levels that were as much as 700-times higher than lagoon release. At 100- percent solids concentration, Ca WET arsenic levels were as much as seven-times higher than the lagoon measured arsenic release. The Ca WET data measured using 100-percent solids concentration demonstrated a much closer correlation with the lagoon release than the 20-percent solids concentration Ca WET results.

TCLP and Ca WET arsenic concentrations in 20-percent solids concentration residuals collected at two and six months are compared with measured arsenic release on the same plots in Figures 4.26 and 4.27. At two months, Ca WET arsenic levels were much higher (by as much as two orders of magnitude) than both TCLP arsenic levels and arsenic levels measured in the lagoon supernatant. TCLP arsenic levels tracked much more closely than Ca WET concentrations with measured lagoon levels. At six months, Ca WET and TCLP concentrations generally increased while lagoon concentrations were generally lower than those measured at two months. These findings are reflected in the ratios shown in Table 4.6.

Table 4.6

Ratio of TCLP and Ca WET to arsenic lagoon release at six months

	Ca WET: Lagoon		TCLP: Lagoon	
Utility	20 percent	Dry	20 percent	Dry
Louisiana Water Co. (New Iberia), LA	442	21.9	18.2	4.1
City of Great Falls, MT	35	6	2.7	0.2
City of Helena, MT	58	48	2.1	1.3
Dept. of Public Utilities (Billings), MT	94	23	0.4	0.1
Lockwood Water Users Assoc. (Lockwood), MT	814	9.0	2.0	0.0
Los Angeles Dept. of Water & Power (LADWP), CA	95	16.1	0.5	0.9
Indiana-American Water Co., IN	1,051	6.6	0.1	0.0

Sand Drying Beds

The results of the bench-scale sand drying bed tests using the LADWP residuals are shown in Figure 4.28. The arsenic concentration in the leachate from each sand bed is shown over three weeks. The arsenic concentrations in the leachates were also measured at four and five weeks at less than the detection limit of the ICP-AES (2-µg/L As). Although the LADWP residuals contained 730 mg/kg arsenic, very little arsenic leached from the sand drying beds. After subtracting the control leachate from the test leachate data, all of the arsenic concentrations were below 10 µg/L.

It was noted during testing that the outside layers of the residuals dried first, creating an impermeable layer to the rainfall water. As the sand drying beds were watered, only the outside layers of the residuals were rinsed. Because the outside layers were dry and hard, it was difficult to get good contact of the residual solids with the rainwater, which significantly reduced the potential for arsenic leaching.

SFBW Clarification

Polymer Testing

A polymer screening study was undertaken to select the two best polymers out of six polymers tested for each SFBW. Screening included two polymers of each charge (cationic, anionic, nonionic) that are commonly used for drinking water treatment. A 1-mg/L dose was added to the SFBW followed by rapid mixing for 30 seconds. To determine the most suitable polymer types, floc formation and settling was noted and the supernatant turbidity was measured after 10 minutes (equivalent to an overflow rate of 0.25 gpm/ft^2). The results of polymer screening are presented in Table 4.7.

Bench-scale SFBW settling tests were conducted using the two best polymers identified during screening. Jar testing (over a range of sedimentation overflow conditions) using six different SFBW samples was performed. The polymer precipitation test results are presented in Figure 4.29. This figure shows the test results using the best polymer dose determined during testing. The polymer doses used to achieve the results noted in the figure are listed in Table 4.8.

Table 4.7

Best polymer types determined from screening SFBW samples

Utility name	Percol LT 22S (cationic)	Percol LT 26 (anionic)	Praestol A3040 LTR (anionic)	Praestol N3100 LTR (nonionic)
New Iberia, LA	✓		✓	
Billings, MT	✓	✓		
Lockwood, MT	✓		✓	
Helena, MT		✓		✓
Great Falls, MT		✓		✓
LADWP, CA	✓		✓	

Table 4.8

Best polymer conditions for treating SFBW

Utility name	Best polymer type	Best Dose (mg/L)	Turbidity (percent removal)	Arsenic (percent removal)
New Iberia, LA	Praestol A3040 LTR	1.0	99	83
Lockwood, MT	Percol LT22S	3.0	74	74
Billings, MT	Percol LT22S	3.0	98	98
Helena, MT	Percol LT26	4.0	99	99
Great Falls, MT	Praestol N3100 LTR	4.0	76	82
LADWP, CA	Praestol A3040 LTR	0.5	99	100

These data are also presented using a comparison of arsenic and turbidity percent removal for polymer conditioning (see Figure 4.30). Results indicate that by using polymer and gravity settling (at an OFR of 0.25 gpm/ft^2) good turbidity removal in the range of 74 to 99 percent was achieved. Also, arsenic was removed at similar percentages as the iron because the arsenic was bound to the suspended solids in the SFBW, and therefore, easily removed by gravity settling. Arsenic removal using polymer ranged from 73 to 100 percent.

Ferric Chloride Testing

The arsenic removal from SFBW samples using the best ferric chloride doses is provided in Figure 4.31. The figure demonstrates that good arsenic removal was achieved using $FeCl_3$, with all decant arsenic concentrations below 8 μg/L. The arsenic percent removals ranged from 90 to 100 percent. Figure 4.32 also shows that ferric chloride can be successfully used to remove turbidity from SFBW, which thereby results in removal of arsenic from solution. Turbidity removals using ferric chloride ranged from 88 to 100 percent. The overall comparison of the coagulant, polymer, and untreated settleability results are presented in Table 4.9.

Table 4.9

Comparison of best SFBW treatments using polymer and ferric chloride

	No chemical treatment		Polymer			Ferric chloride		
Utility name	As remaining (μg/L)	As (percent removal)	Dose (mg/L)	As remaining (μg/L)	As (percent removal)	Dose (mg/L)	As remaining (μg/L)	As (percent removal)
New Iberia, LA	30	41	1.0	7	83	120	5	90
Great Falls, MT	29	9	4.0	6	82	120	BDL	100
Helena, MT	159	3	4.0	BDL	99	20	3	98
Billings, MT	24	55	3.0	BDL	98	40	BDL	99
Lockwood, MT	10	25	3.0	4	74	40	BDL	96
LADWP, CA	50	73	0.5	BDL	100	20	7	96

BDL = Below Detection Limit (< 2 μg/L)

These data clearly indicate that chemical conditioning (using either a polymer or coagulant) is necessary to remove arsenic from solution, at a conventional settling overflow rate of 0.25 gpm/ft^2, to a concentration less than 10 μg/L. Gravity settling alone (no chemical addition), however, may be able decrease arsenic to below 50 μg/L for direct discharge applications.

Because both ferric chloride and polymer were equally effective for removing arsenic, utilities should focus on polymer conditioning as their first option for treating SFBW when chemical addition is required, due to lower additional solids generation from conditioning.

TCLP arsenic concentrations were measured for each of the SFBW solids samples collected from the best chemical conditioning jar test, with the exception of the New Iberia solids. The TCLP arsenic concentrations are listed in Table 4.10.

Table 4.10

TCLP arsenic for SFBW solids

Utility name	Best treatment condition	TCLP arsenic (mg/L)
Great Falls, MT	80-mg/L Ferric chloride	0.002
Helena, MT	4.0-mg/L Percol LT–26 Polymer	0.008
Billings, MT	3.0-mg/L Percol LT-22s Polymer	0.003
Lockwood, MT	40-mg/L Ferric chloride	0.002
LADWP, CA	0.5-mg/L Praestol A3040 LTR Polymer	0.018

These data demonstrate that leaching of arsenic from the settled solids using the TCLP test did not occur. None of these tests resulted in arsenic concentrations even within two orders of magnitude of the regulatory limit of 5.0 mg/L.

Lime Conditioning Study

General

Lime may be used as a conditioning chemical in the dewatering of coagulant residuals, especially in conjunction with filter-press dewatering. Because arsenic solubility may be impacted by pH changes and by the concentration of calcium in solution, a series of tests was conducted to assess the impact of lime and separately to evaluate the role of calcium in the leaching of arsenic from residuals.

Chemical Conditioning Tests

Caustic soda, soda ash, and lime were each evaluated to assess their effect on arsenic leachability from ferric and alum residuals over a range of pH values.

Caustic Soda

Each residual was treated with caustic soda over the pH range encountered with lime treatment. This test was performed to determine whether the observed arsenic release was solely due to a pH effect, i.e., hydroxide concentration. Total arsenic increased dramatically as the pH was increased with caustic soda (Figure 4.33). The same trends were observed for both the ferric and aluminum residuals. As much as 30 percent of the available arsenic was leached from the ferric residuals at pH 11.6 when no lime (calcium) was present. This trend is consistent with the expectation based on surface complexation (Dzombak et al. 1990);

$$FeOH + AsO_4^{3-} + 3H^+ \leftrightarrow FeH_2AsO_4 + H_2O \quad (4.1)$$

As the pH is raised, the hydrogen ion concentration decreases, shifting equilibrium toward the reactants (left side of the equation) and releasing arsenate to solution.

Soda Ash

Each residual was also treated with soda ash over the pH range encountered with lime treatment. This test was performed to determine the combined effect of pH and carbonate on arsenate release. The same trends were observed as with the caustic soda, and in a few cases, the extra carbonate increased arsenic leaching. A maximum pH of 10.3 was obtained with soda ash using both ferric and alum residuals. Total arsenic increased dramatically as the pH was increased (Figure 4.33). As much as 12 percent of the available arsenic was leached from the ferric residuals at pH 10.3 when soda ash was present. This phenomenon also is consistent with the expectation based on surface complexation (Villalobos and Leckie 2001);

$$FeOH + CO_3^{2-} + 2H^+ \leftrightarrow FeOCOOH + H_2O \quad (4.2)$$

An increase in the carbonate concentration results in an equilibrium shift towards FeOCOOH. Therefore, there is an increase in the total number of surface sites on the ferric hydroxide that are occupied by carbonate. This increase, coupled with competition from hydroxide at high pH (from Equation 1), is the reason that the total arsenic concentrations are highest at pH 10.3 when soda ash is used.

Lime Dewatering

The data from the lime addition tests yielded some surprising results. For the ferric residuals, the arsenic leachate concentration appeared to peak at a lime dose of approximately five percent or a pH of 11 (Figure 4.33) and at a level of only 0.55 percent arsenic leached. By comparison, caustic soda and soda ash addition resulted in a much higher total arsenic concentration as pH increased. The lime data show that the proposed theory, more total arsenic leached with increasing pH, is incorrect. This appears to be attributable to the increase in calcium in the water associated with the use of lime, as will be discussed later. Other data from the ferric residuals testing are summarized in Table 4.11.

Table 4.11

Lime effects on ferric residuals

		Leachate concentrations				
Lime percent (dry w/w)	pH	As (µg/L)	Iron (mg/L)	Aluminum (mg/L)	Calcium (mg/L)	Percent As leached
0.0	7.5	10	0.004	0.017	3.7	0.06
1.4	9.1	42	0.014	0.077	23	0.29
4.4	10.8	82	0.003	0.268	47	0.55
9.1	12.1	29	0.002	0.915	291	0.20
18.5	12.6	21	0.0	1.252	500	0.15

Table 4.12

Lime effects on alum residuals

Lime percent (dry w/w)	pH	As (µg/L)	Iron (mg/L)	Aluminum (mg/L)	Calcium (mg/L)	Percent As leached
		Leachate concentrations				
0.0	7.1	5	ND	0.15	27	0.04
1.4	8.1	6	ND	0.62	39	0.05
5.4	9.3	5	ND	7.80	30	0.04
9.8	9.8	5	ND	17	38	0.03
19.6	10.2	17	ND	93	117	0.13

The alum residuals treated with lime showed a constant level of leaching and an abrupt increase in soluble arsenic at a lime dose of 20 percent or a pH of 10 (Figure 4.34). This phenomenon can be explained by looking at the solubility of aluminum. At high pH values the solubility of Al increases dramatically. This is borne out by the data from this experiment. At a pH of 10.2 the soluble Al concentration rose to 93 mg/L. At this pH approximately four percent of the total Al was in the soluble form (Table 4.12). Since the arsenic was initially sorbed to the alum solids, it can be expected that the soluble arsenic concentration will also increase as more aluminum is dissolved.

TCLP and California WET results are shown in Table 4.13. The amount of lime dosed did not seem to affect the final arsenic concentrations. At very high lime doses (20-percent calcium) there was a small decrease in the arsenic concentration. The ferric residuals results ranged from 0.006 to 0.018 mg/L for the TCLP and 10.9 to 14.3 mg/L for the Ca WET. The alum residuals results ranged from 0.024 to 0.070 mg/L for the TCLP and 6.5 to 8.8 mg/L for the Ca WET. Since the regulatory limit is 5.0 mg/L for both the TCLP and Ca WET procedures, each residual passed the TCLP test but failed the Ca WET test for all lime doses.

Table 4.13

Lime effects on coagulation residuals: TCLP and Ca WET data

Lime percent (dry w/w)	pH	TCLP (mg/L As)	Ca WET (mg/L As)
Ferric coagulation			
0.0	7.5	0.006	11.6
1.4	9.1	0.006	14.3
4.4	10.8	0.018	13.6
9.1	12.1	0.014	13.1
18.5	12.6	0.010	10.9
Alum coagulation			
0.0	7.1	0.031	8.8
1.4	8.1	0.070	7.7
5.4	9.3	0.050	7.9
9.8	9.8	0.037	8.0
19.6	10.2	0.024	6.5

Possible Reasons for Reduced Arsenic Leaching in the Presence of Calcium

Two hypotheses were formulated as to why soluble arsenic decreases with increasing pH in the presence of calcium, as opposed to when only hydroxide and/or carbonate is present. The first hypothesis is that arsenic, not bound to the solids, is reacting with the calcium to form a calcium arsenate solid (Bothe and Brown 1999). The second theory is that calcium is somehow enhancing the surface adsorption of arsenic onto the solids in solution. Each of these hypotheses was investigated in turn.

Calcium Arsenate Solid Formation

Formation of calcium arsenate solids has been observed by some researchers at high pH (Bothe and Brown 1999). To examine whether formation of this solid could explain the results of this work, an experiment was conducted with 825 mg/L calcium and 2,000 μg/L arsenic in solution with no residuals present, using the pH range encountered in the previous chemical conditioning tests. Figure 4.35 demonstrates that a calcium arsenate solid did not form at a pH below 11.5. At pH 12 and pH 12.5 a calcium arsenate solid may be forming and/or the arsenic may be sorbing to lime that did not dissolve. The calcium and arsenic are not precipitating from solution in a stoichiometric amount, therefore, more than one mechanism could be responsible.

Divalent Cation Theory

Since there was no calcium arsenate solid forming up to pH 12, the next hypothesis was that the calcium must have somehow been affecting the surface properties of the ferric solids. As the pH increases the surface charge of ferric or alum solids will become more negative. Since arsenic is naturally present in anionic form (i.e., arsenate or arsenite), there would be a repulsion between the arsenic and the ferric hydroxide solids, resulting in less sorption. It appears that the calcium is neutralizing this negative surface charge, and the arsenic remains sorbed to the ferric hydroxide solids.

To test this theory, various amounts of calcium (as calcium chloride, not lime) or magnesium was dosed with a fixed amount of ferric solids. Magnesium was used as another representative divalent cation, but one which could not form calcium arsenate. Figure 4.36 shows the results obtained when the pH was held at 10. There is very good agreement between the amount of arsenic leached versus the molar amount of either calcium or Mg remaining in solution. The additions of either calcium or Mg yielded a similar outcome. The same test was also conducted at pH 11. Results for calcium are shown in Figure 4.36, however, very little Mg could be kept in solution due to the insolubility of magnesium hydroxide and as a result, the Mg data are not provided. The trend is similar to that observed at pH 10.

Having established that divalent cations enhance the arsenic sorption to ferric hydroxide residuals at high pH, an explanation for this result was sought. The first idea is that cations can reduce the electrostatic repulsion for negatively charged arsenate and permit attachment to the surface. The second theory is that the divalent cation actually sorbs to the ferric hydroxide surface (Dzombak et al. 1990);

$$FeOH + Ca^{2+} \leftrightarrow FeOHCa^{2+} \quad (4.3)$$

$$FeOH + Ca^{2+} \leftrightarrow FeOCa^{+} + H^{+} \quad (4.4)$$

It has been the researcher's experience that the benefits of calcium occur above pH 7.8. This has ramifications for hard water systems since pH will be above this.

The sorbed calcium provides more positively charged surface sites (i.e., a less negatively charged surface) to which negatively charged arsenate might sorb. It has been shown that calcium adsorbs to the ferric hydroxide surface in appreciable amounts at high pH values (Smith and Edwards 2001). A mass balance on calcium was performed for each lime dose experiment, and the amount of particulate calcium was determined. The ratio of particulate calcium to iron as a function of pH is graphically shown in Figure 4.37 and indicates considerable partitioning of calcium to the solid phase. Since a molar ratio of 0.25 is considered to be the maximum amount of calcium that can sorb to ferric hydroxide surface, it is readily apparent that some of the solid phase calcium must be present above pH 11 (Dzombak et al. 1990).

Diffuse Layer Compression

A model based on Chen was used to evaluate the data at both pH 10 and 11 Figure 4.38 (Chen 2001). This model utilizes parameters such as ionic strength, pH, total arsenic, and total iron to estimate the percentage of arsenic that will be sorbed to a fresh iron surface. The role of calcium in the model is to shield repulsive forces between sorbed arsenic, and no consideration is given to sorbed calcium species. The goal was to determine the extent to which trends in the data could be predicted even though it was expected that the quantitative accuracy would be limited for the aged iron solids present in this work. The modeling proceeded as follows. First, the model was calibrated by selecting a total quantity of iron surface sites available for the experiment in which no lime was added so that the model prediction of arsenic sorption matched the initial experimental value. The calibration was needed because only amorphous iron hydroxide has surface sites available for sorption, and no procedure was available to quantify this amount. Soluble calcium concentrations in each experiment were entered into the model to obtain a prediction of the change in arsenic sorption. As one can see from the figure, there is very good agreement between the actual data and the model at these test conditions. The model, however, does not predict or explain the increased arsenic sorption at higher pH values (above 11). As noted earlier, there was evidence that a different calcium solid phase was forming at these pH values, and it is possible that arsenic was sorbing to these solids.

During this portion of the study no speciation of arsenic was performed and, therefore, it is impossible to ascertain whether the soluble arsenic was present as arsenate or arsenite. It was assumed in the above modeling that all arsenic was present as arsenate (As (V)) since no effort was made to keep the residuals in an anaerobic or anoxic environment. Since the residuals were of unknown age however, some reduction of arsenate could have taken place over time. Figure 4.9 illustrates modeling results that show both the speciation of arsenic and the amount of calcium in solution as critical in determining sorption efficiency. The results indicate that increasing calcium content in the ferric hydroxide/water solution increases arsenate sorption but decreases arsenite sorption. This phenomenon could explain the modeling results described above and shown in Figure 4.38. If a portion of the arsenic were present as arsenite the model would predict less overall sorption resulting in an upward shift in the model curves shown in the figure. This shift would allow the model to provide a better fit to the actual data at higher levels of calcium.

QA/QC Results

Arsenic concentrations in all blanks were below the instrument's detectable limit. Matrix 'spikes and recoveries' were performed by adding 220-mg/L arsenic to the sample. Recoveries ranged between 85 and 115 percent in most cases, indicating little matrix interference.

Several replicates of the chemical conditioning tests were performed during the course of this study. Problems with QA/QC on these tests, however, precluded the use of that data in this report.

Even though these data were not used, they still showed the same general trends that are described in this report.

Lime Conditioning Summary

Calcium addition in the form of lime appears to be a practical and viable method for immobilizing arsenic sorbed to ferric hydroxide and aluminum hydroxide residual solids. Ferric residuals appear to retain their sorbed arsenic to a greater extent than alum residuals at the higher pH values encountered with lime use, probably due to the increased solubility of aluminum at these high pH values. Surface complexation modeling also appears to be good method to determine the appropriate level of soluble calcium required to minimize arsenic release.

CONCLUSIONS

Lagoon Simulations

The lagoon storage of arsenic containing residuals has different effects on arsenic release. Residuals stored at higher percent solids exhibit higher total arsenic concentrations. The release of arsenic in ferric chloride residuals is directly related to the change in total iron. The release of arsenic in aluminum hydroxide residuals might be related to the amount of ambient iron present or possibly other unknown factors. Softening residuals released less than 6-ppb arsenic due to high redox potentials and a low concentration of arsenic. It should also be noted that presence of biodegradable organic matter increases the arsenic solubility, most likely due to biological activity that results in anaerobic conditions.

As redox potentials decrease, total arsenic increases. Total iron also increases with decreasing redox potentials. The storage of residuals in lagoons creates reduced environments where arsenic will be released, possibly to surface and groundwater. Lagoon storage of arsenic -containing wastes should be more strictly regulated in the future to prevent contamination.

TCLP and California WET results are affected by lagoon storage of the residuals. As residuals age, changes in the redox potentials influence changes in the total arsenic concentrations.

The amount of total arsenic present influences the results of the toxicity tests. Neither of the toxicity tests takes into account the changes that occur in redox potential.

Sand Drying Beds

The leachate of residuals on bench-scale sand drying beds contained less than 10-µg/L As. The residuals air dry on the sand beds under oxidized conditions, therefore, reducing conditions that could cause arsenic release are not likely to occur.

Lime Dewatering Conclusions

The following conclusions were developed from the lime dewatering investigation. These findings are as follows:

- Arsenic release from ferric hydroxide residuals increases with increasing pH in the absence of calcium
- Arsenic release from alum residuals also increases with increasing pH even in the presence of calcium. Calcium reduces the effect when compared to increasing the pH with either caustic soda or soda ash
- Lime has little effect on the TCLP or California WET values for both ferric and alum residuals
- The reduction in arsenic leachability at higher pH values is most likely due to the divalent cation effect of calcium and not to the formation of a calcium arsenate solid

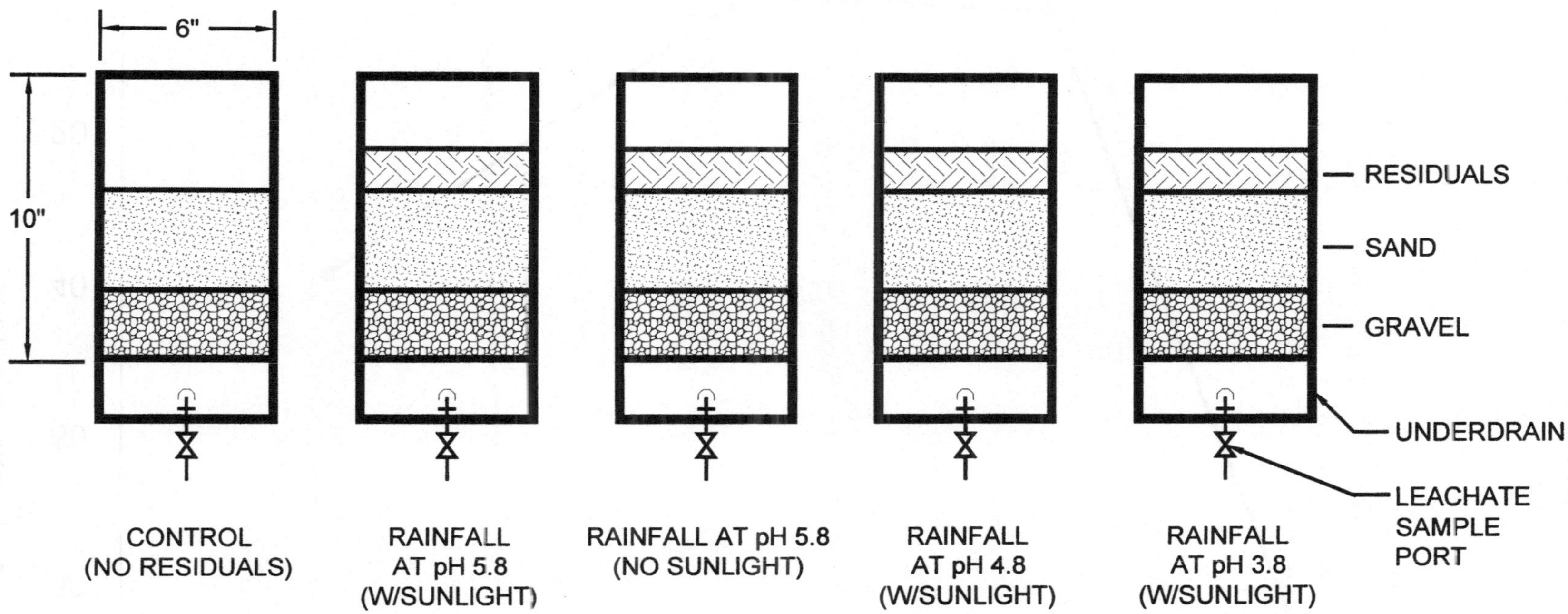

Figure 4.1 Pilot-scale sand drying bed test cylinders

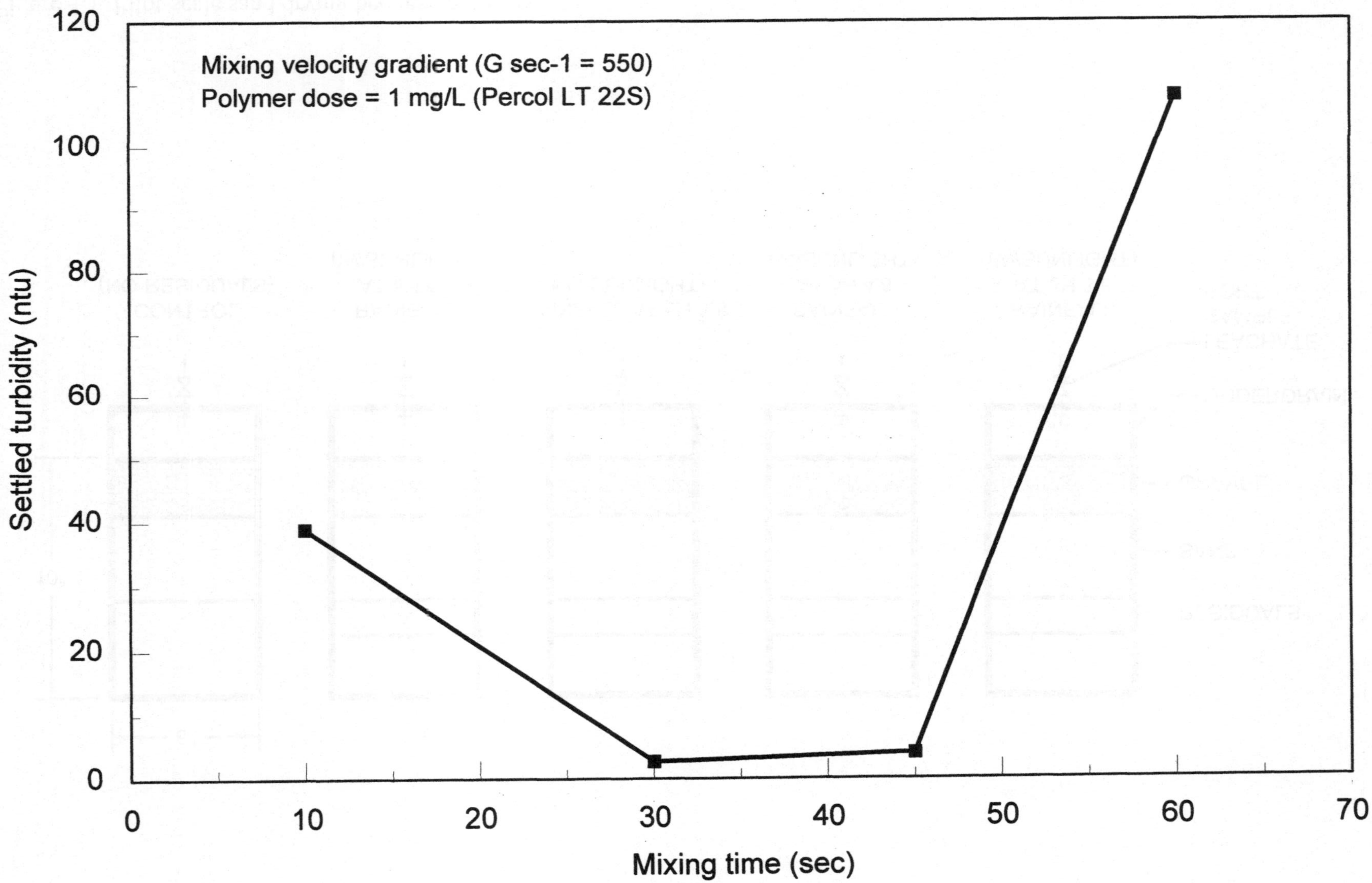

Figure 4.2 Impact of mixing conditions on settled turbidity using New Iberia SFBW

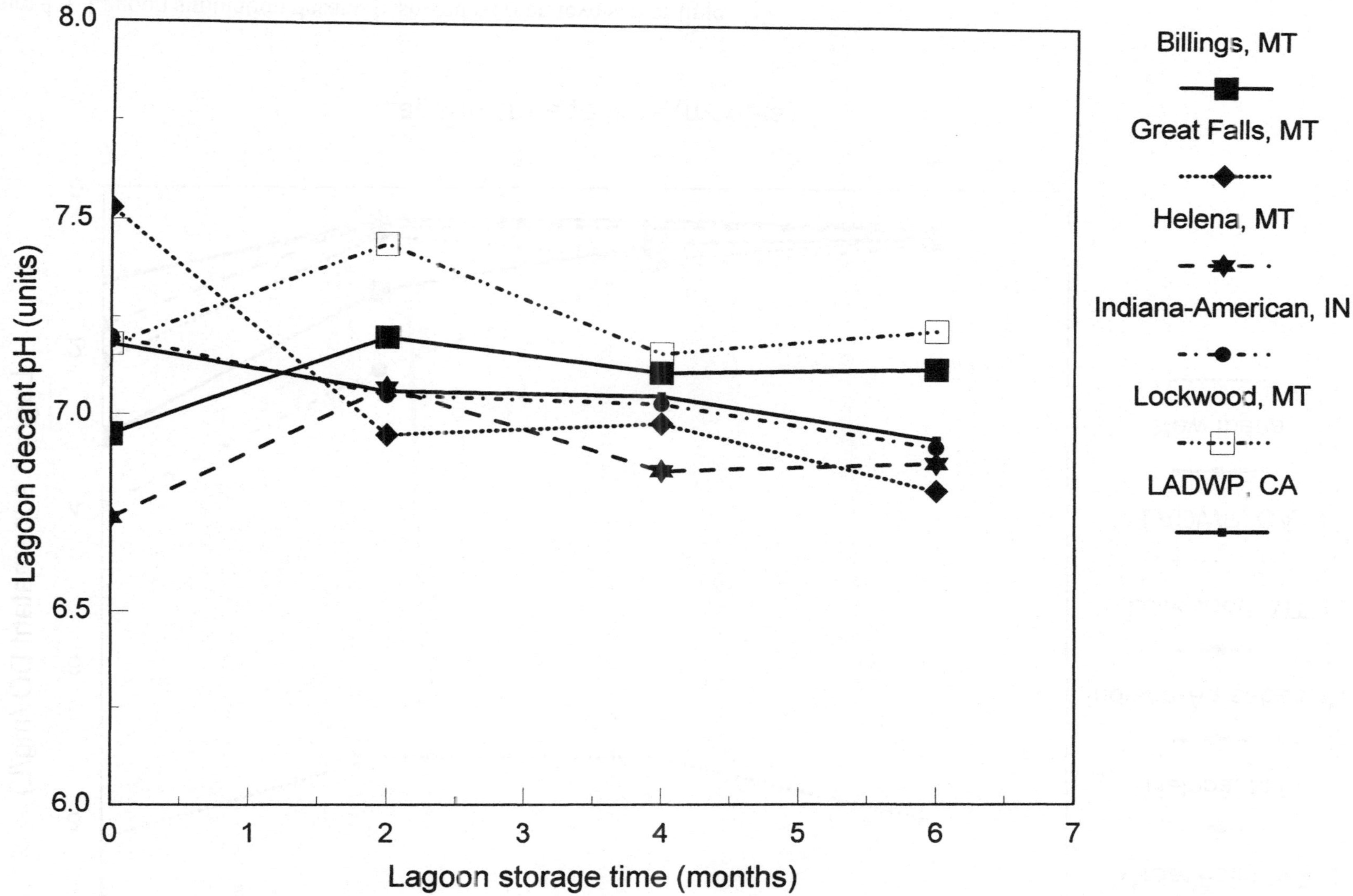

Note: New Iberia pH remained at 11.6 during lagoon simulation testing

Figure 4.3 Lagoon simulation decant pH over time

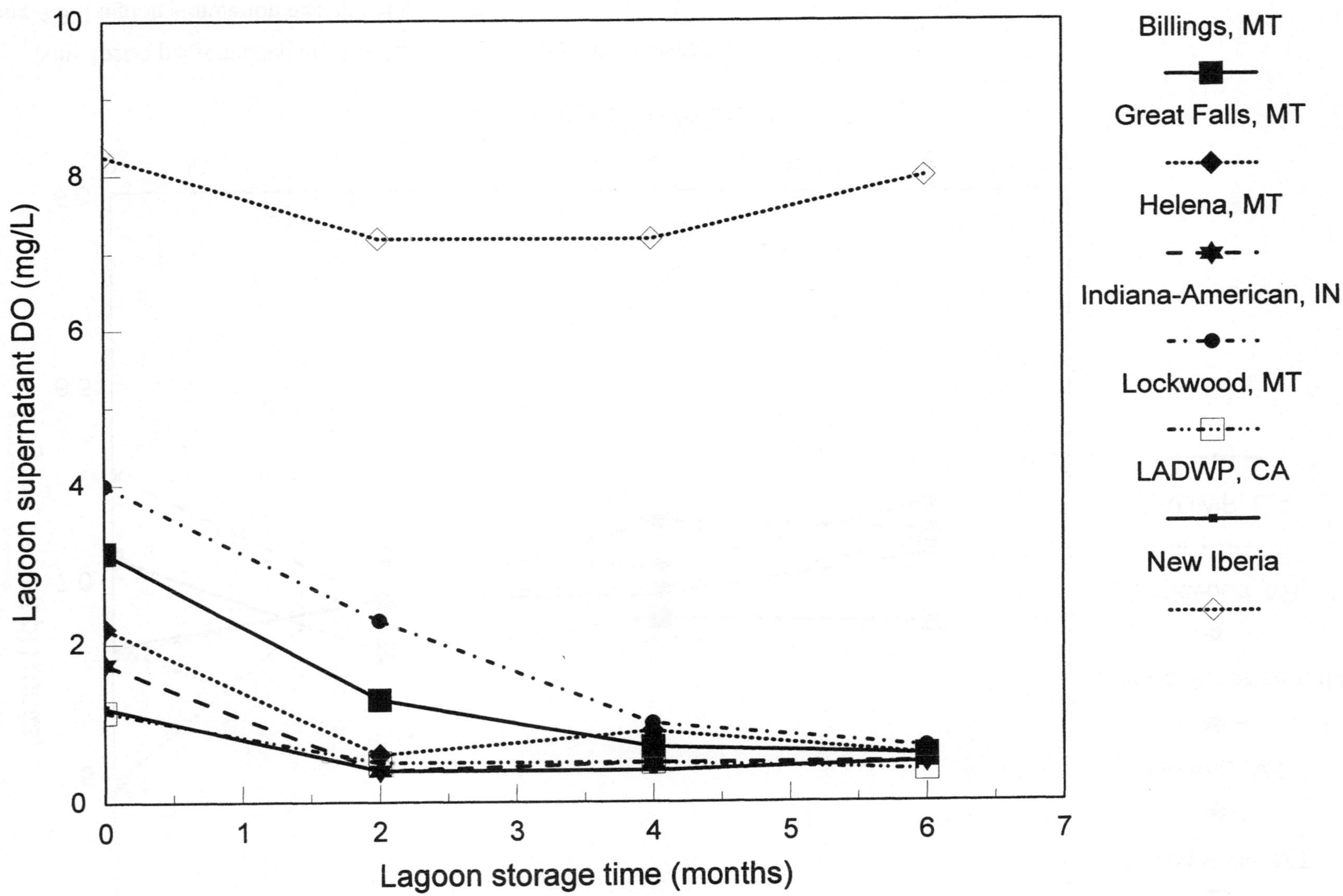

Figure 4.4 Lagoon simulation decant dissolved oxygen levels over time

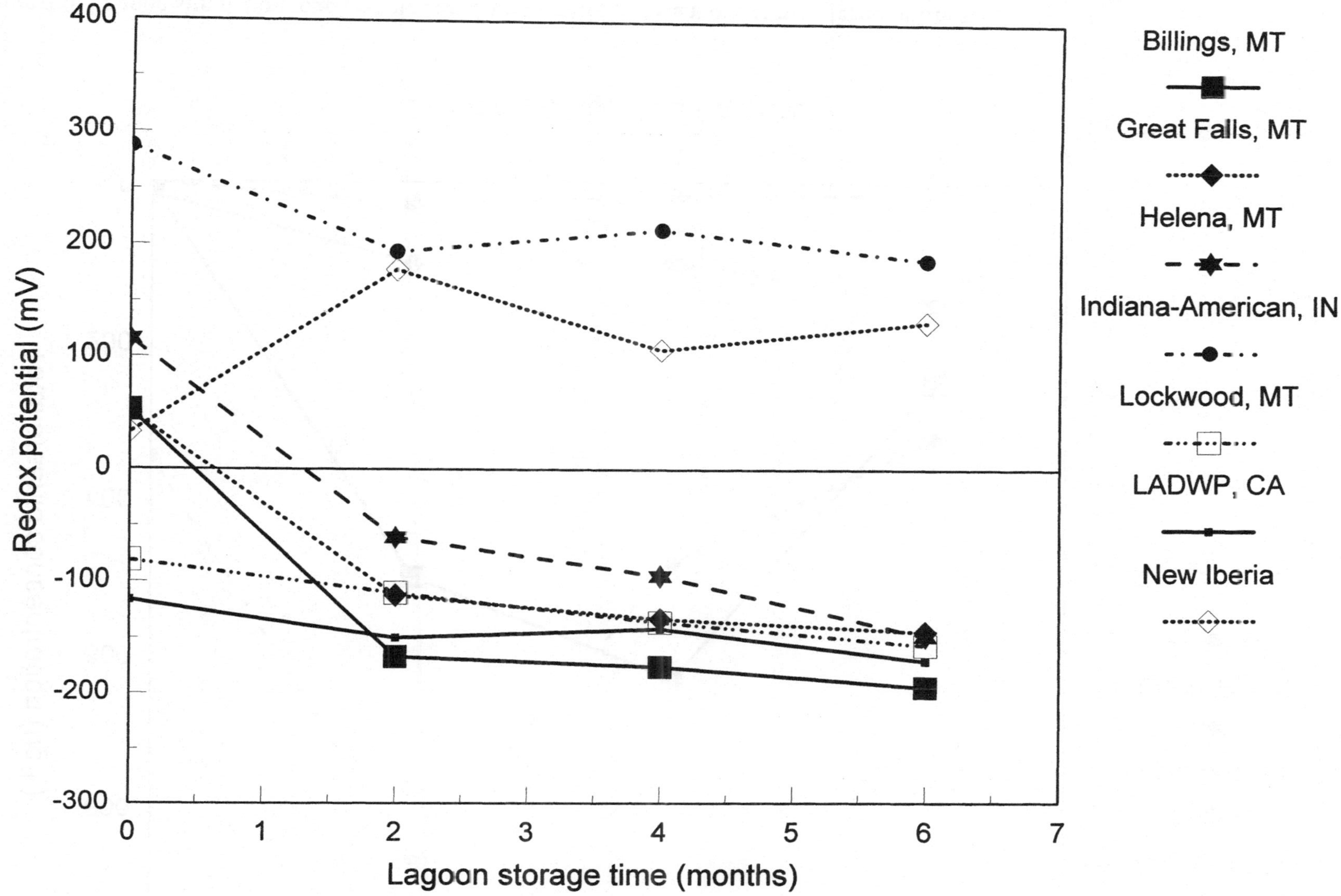

Figure 4.5 Lagoon simulation redox potential over time

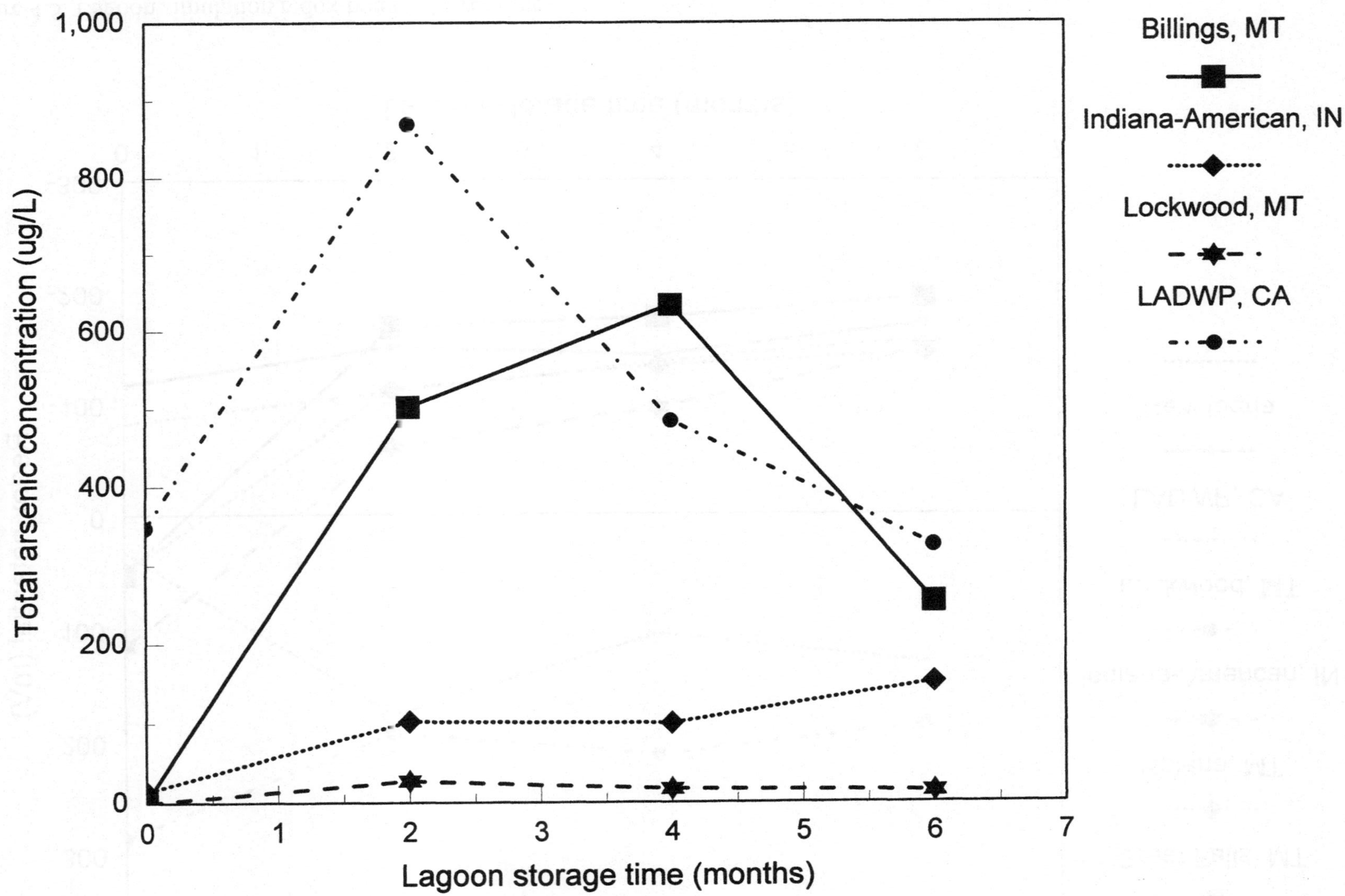

Figure 4.6 Lagoon simulation total arsenic concentration in the decant over time: ferric utilities

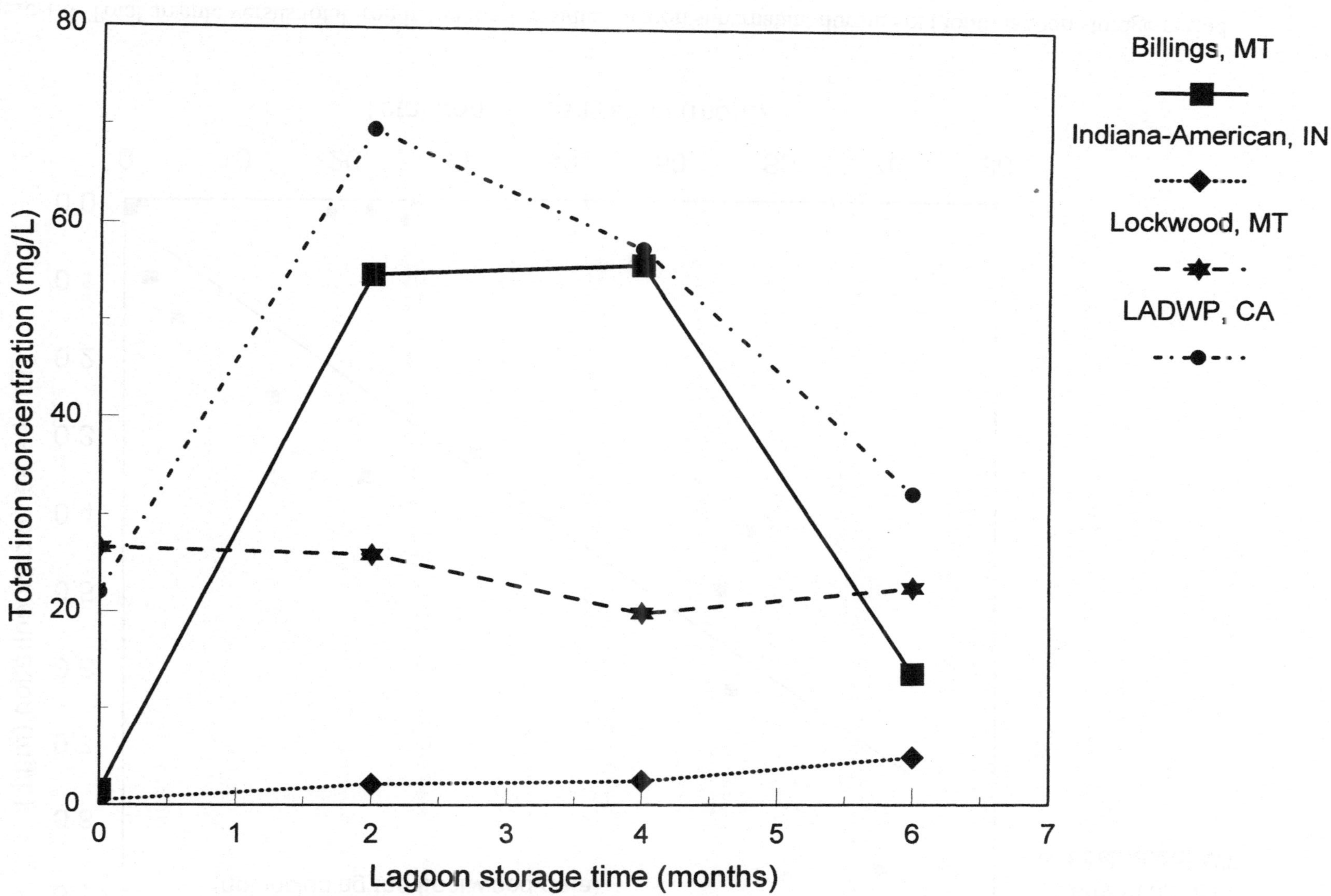

Figure 4.7 Lagoon simulation total iron concentration in the decant over time: ferric utilities

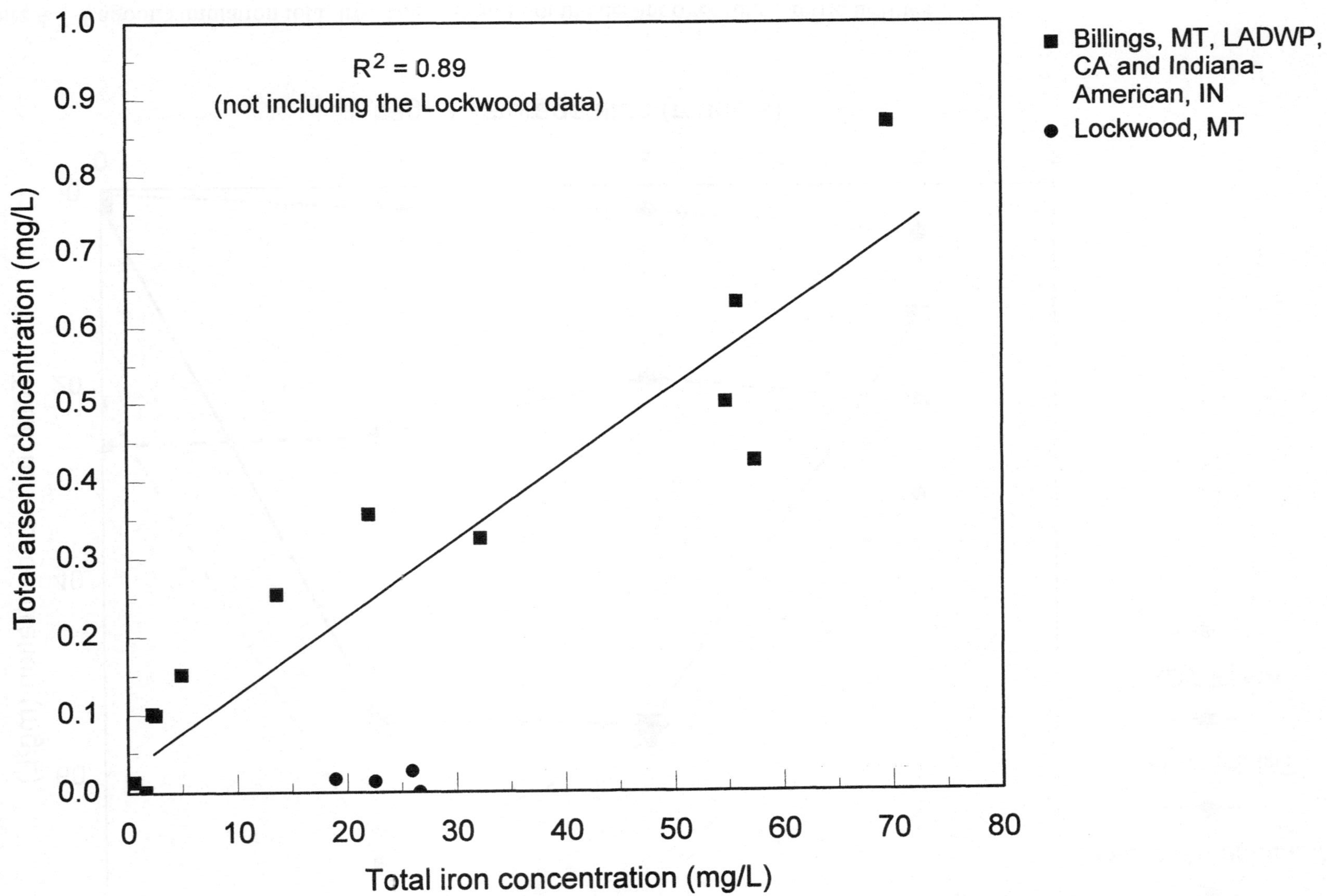

Figure 4.8 Total arsenic versus total iron in the ferric residual lagoon supernatant during six month lagoon storage period

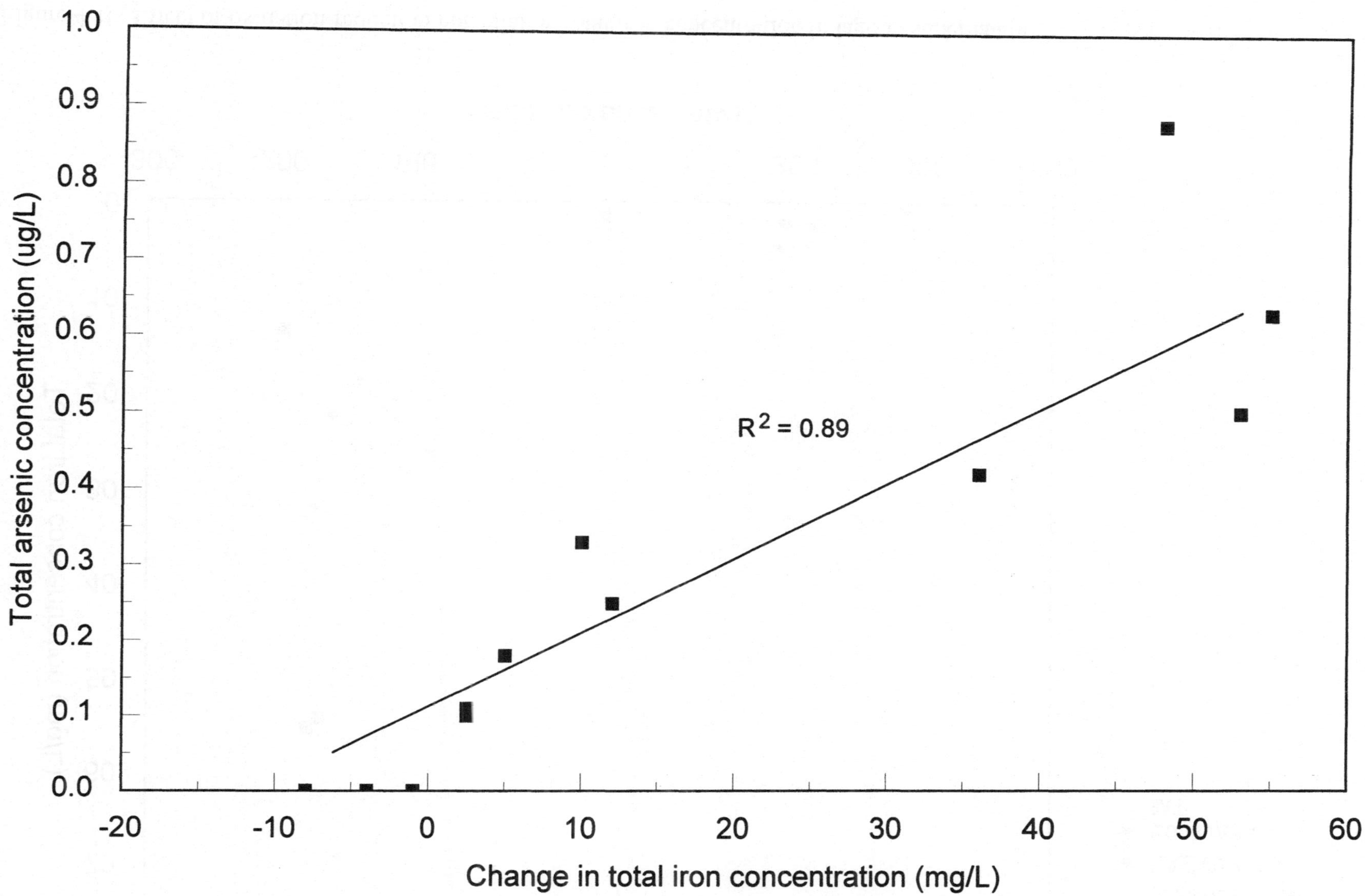

Figure 4.9 Arsenic versus change in iron in the ferric residual lagoon supernatant during six month lagoon storage period

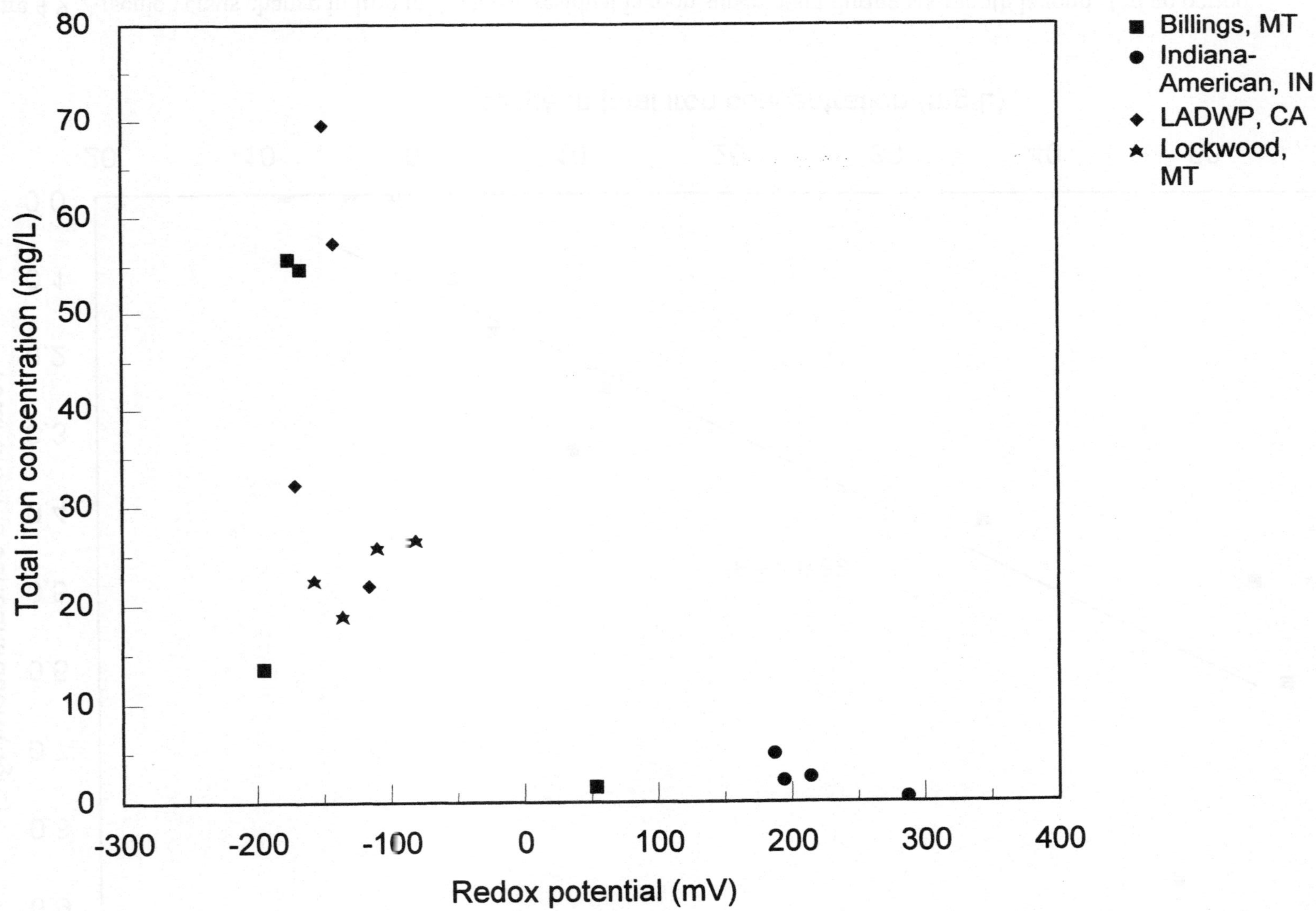

Figure 4.10 Effect of oxidation-reduction potential on total iron concentration in lagoon supernatant

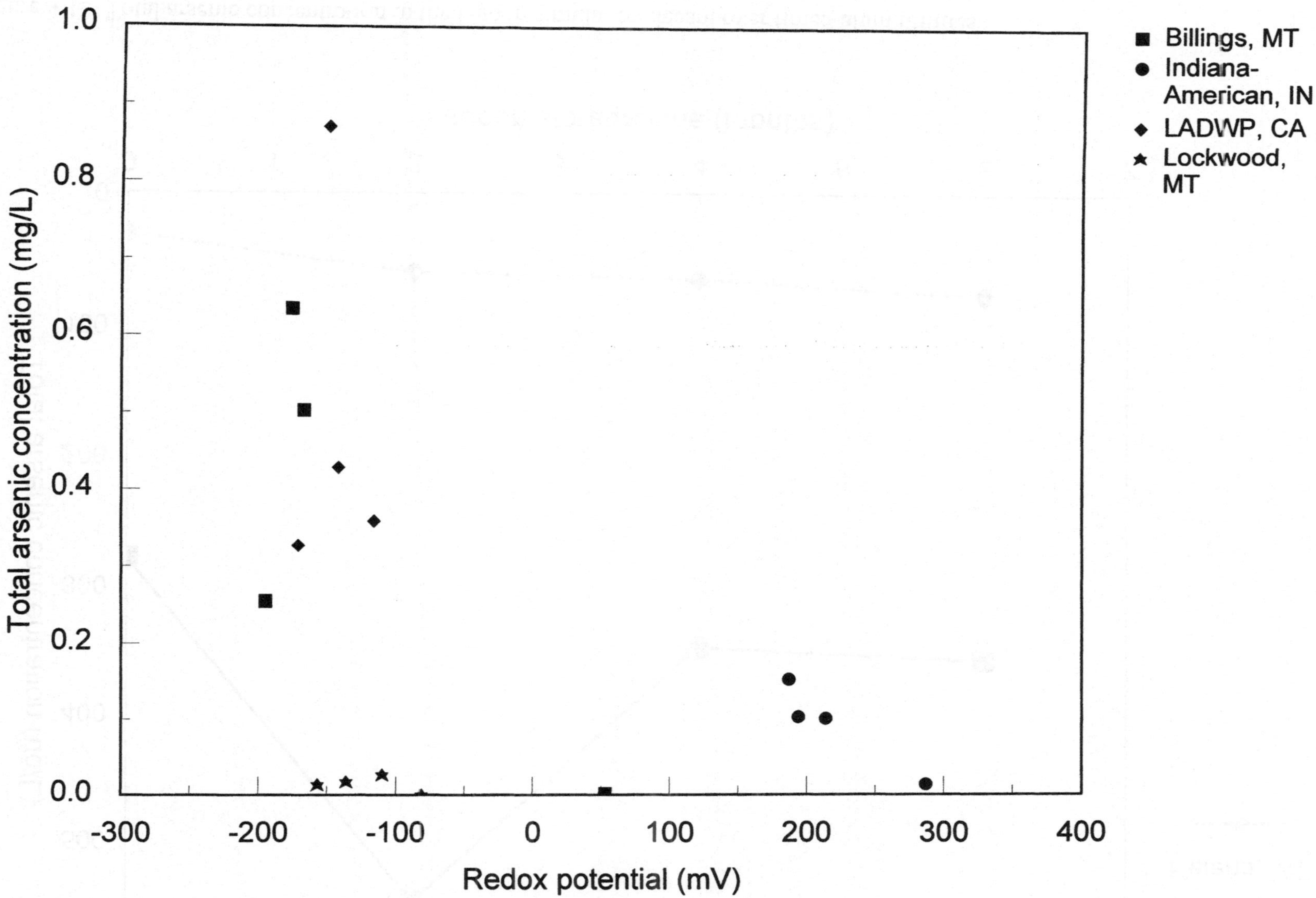

Figure 4.11 Effect of oxidation-reduction potential on total arsenic concentration in lagoon supernatant

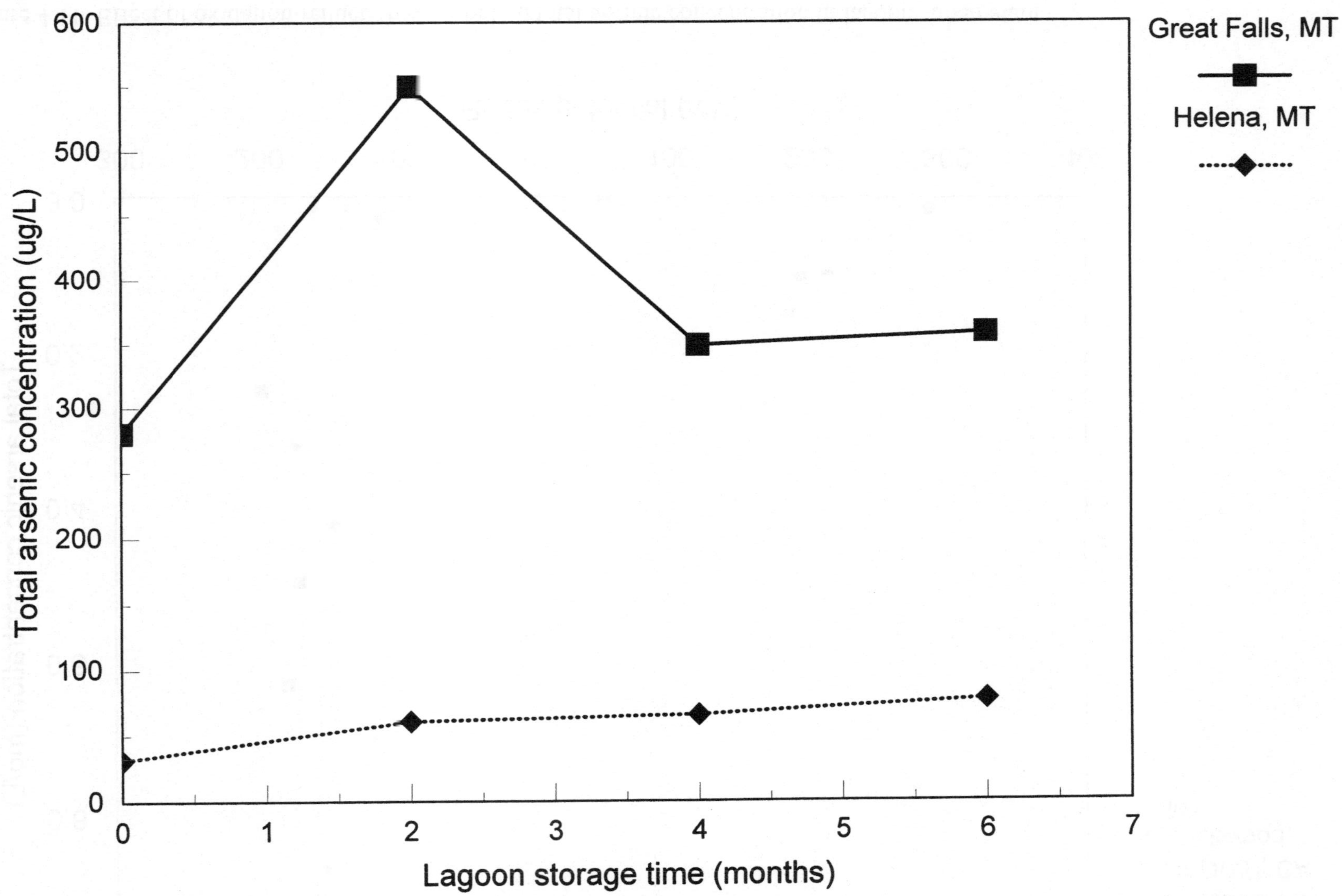

Figure 4.12 Total arsenic concentration in the lagoon simulation decant over time: alum utilities

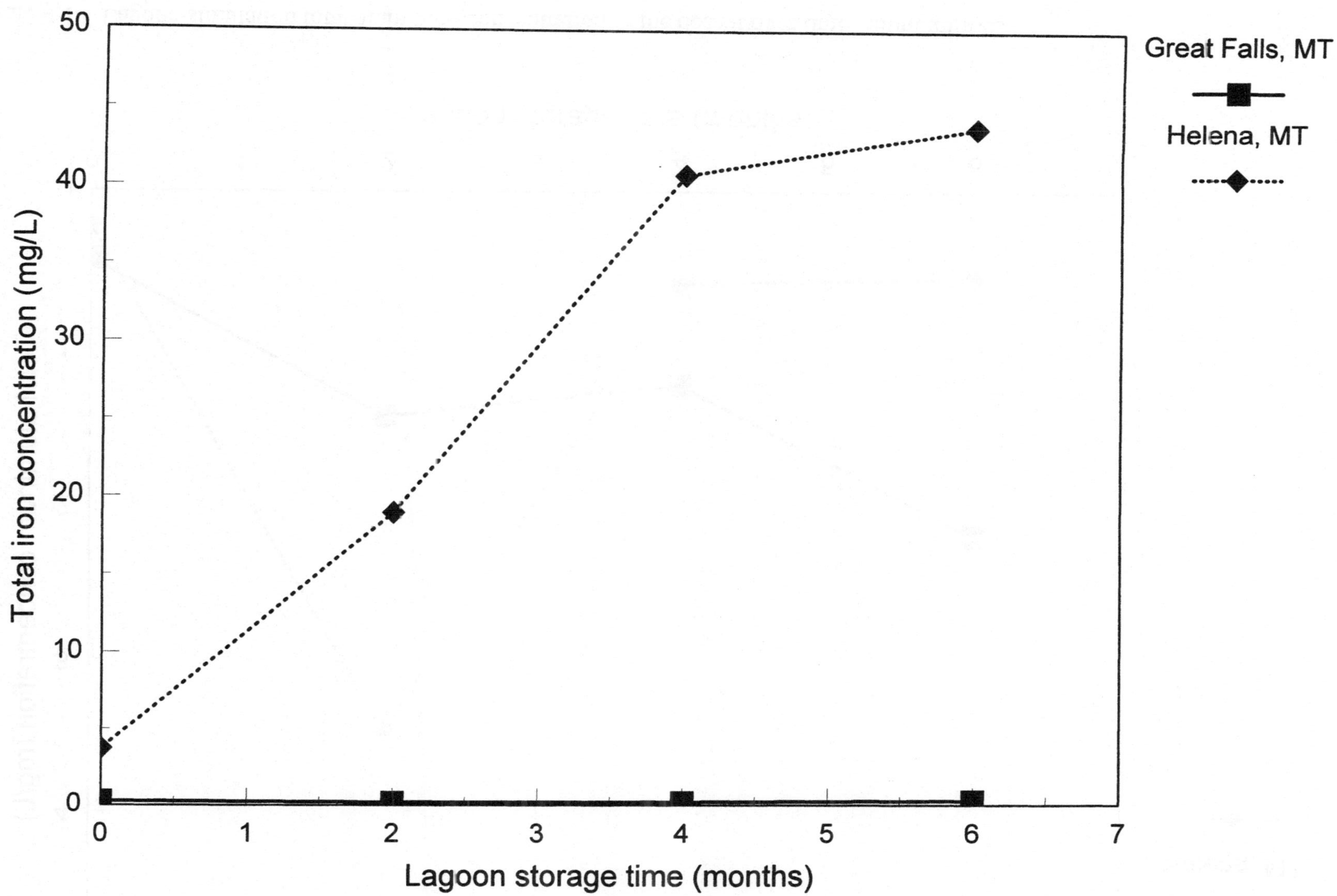

Figure 4.13 Lagoon simulation total iron concentration in the decant over time: alum utilities

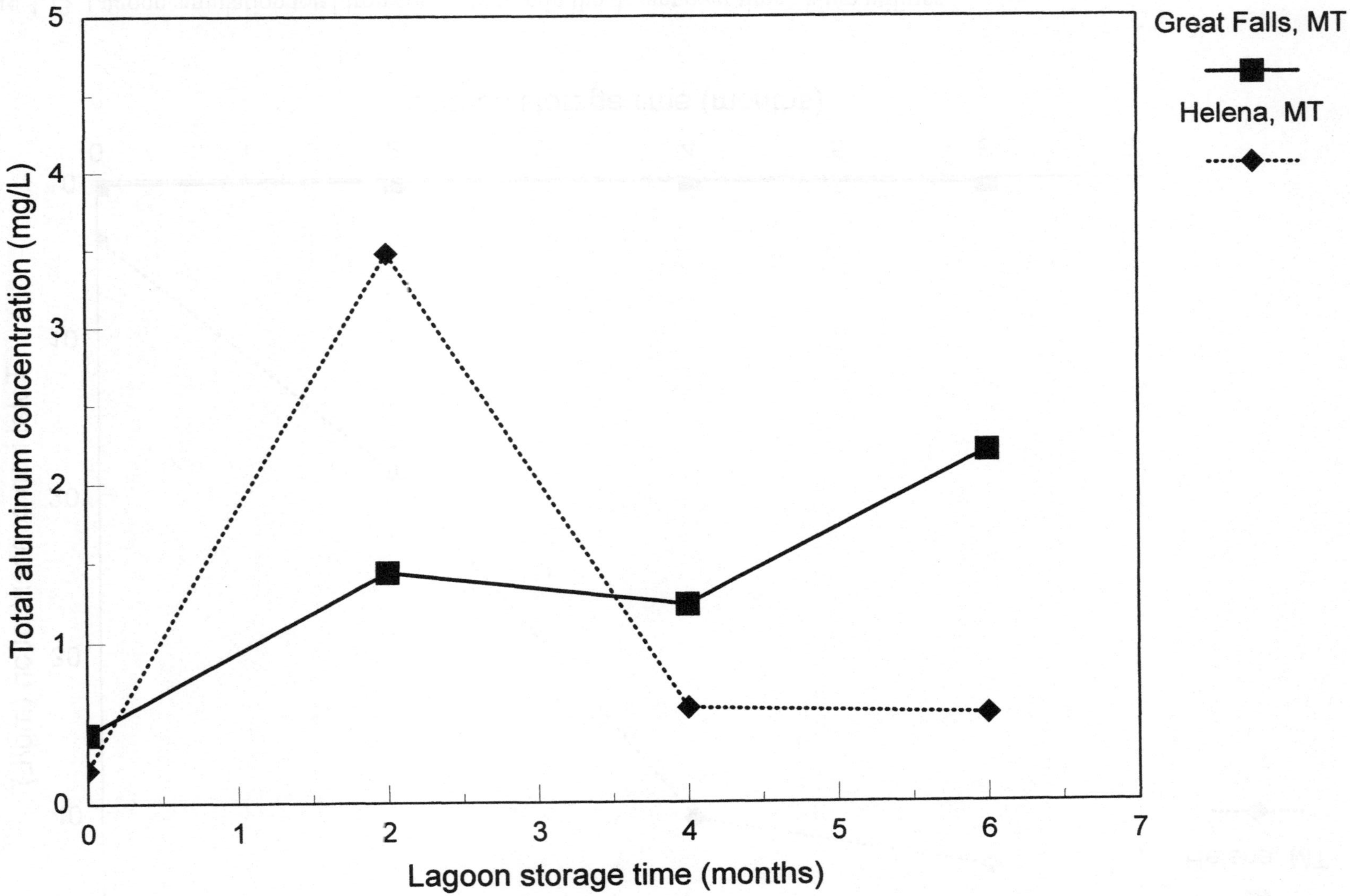

Figure 4.14 Lagoon simulation total aluminum concentration in the decant over time: alum utilities

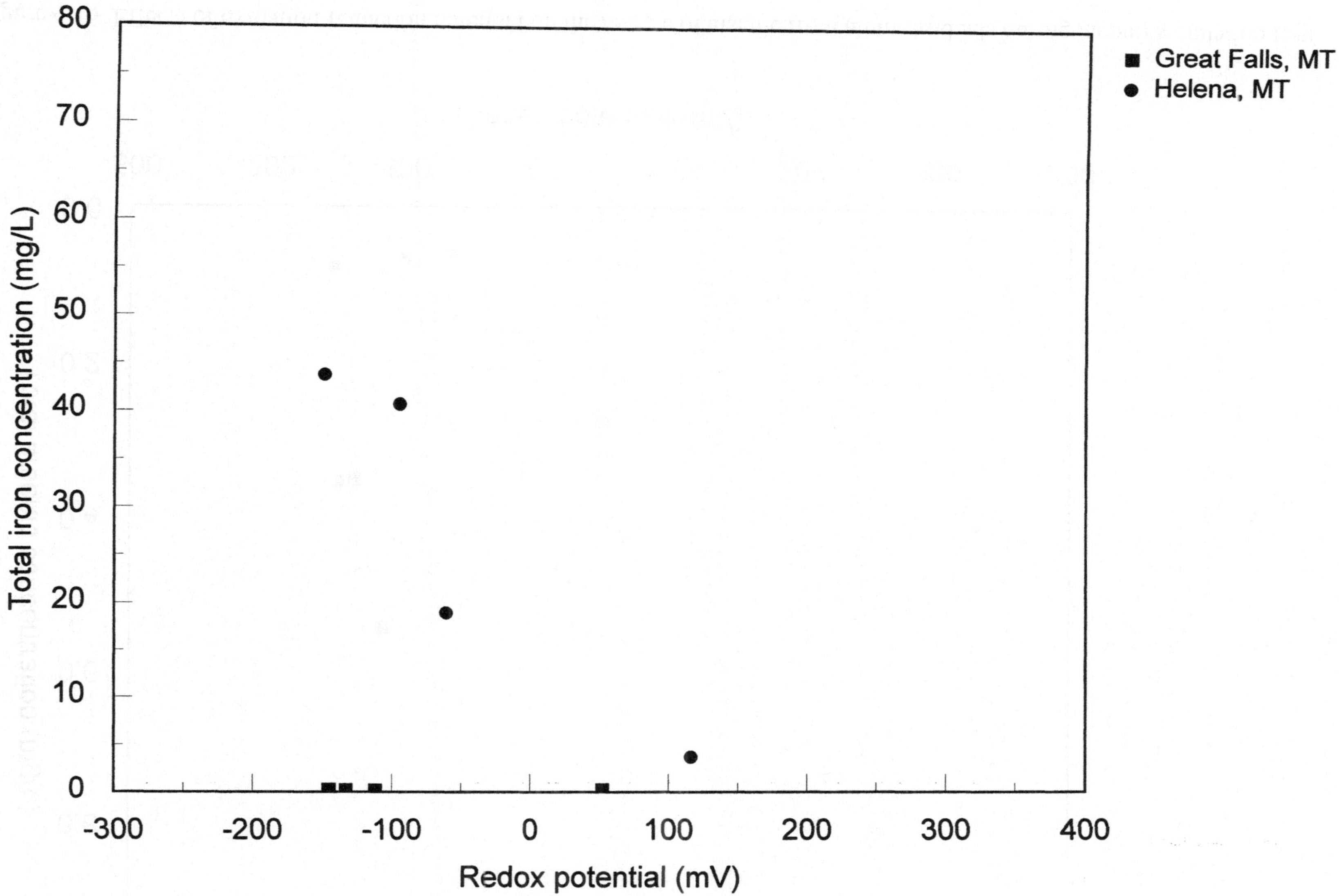

Figure 4.15 Effects of oxidation-reduction potential on the release of iron from alum residuals during lagoon simulation tests

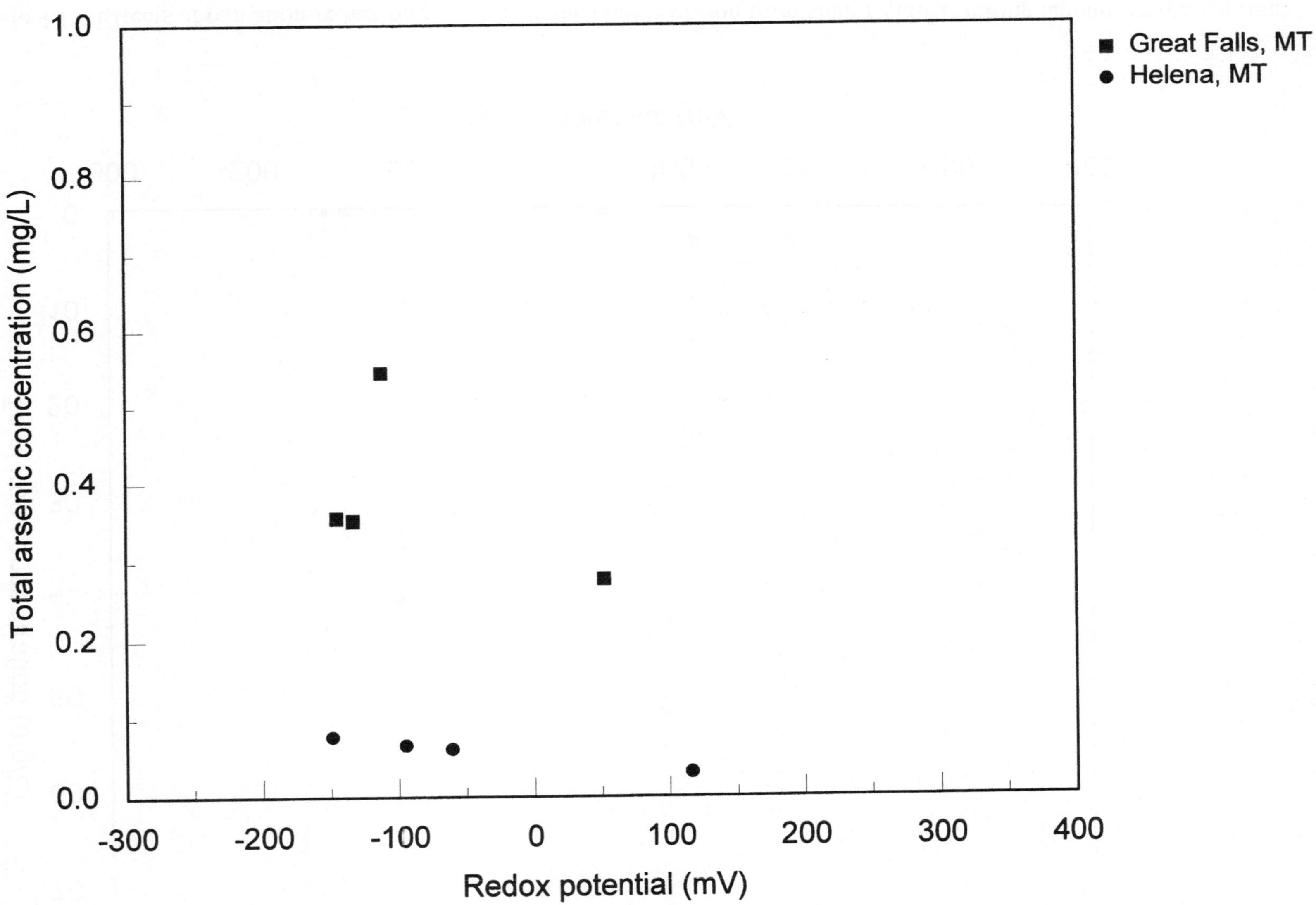

Figure 4.16 Effects of oxidation-reduction potential on the release of arsenic from alum residuals during lagoon simulation tests

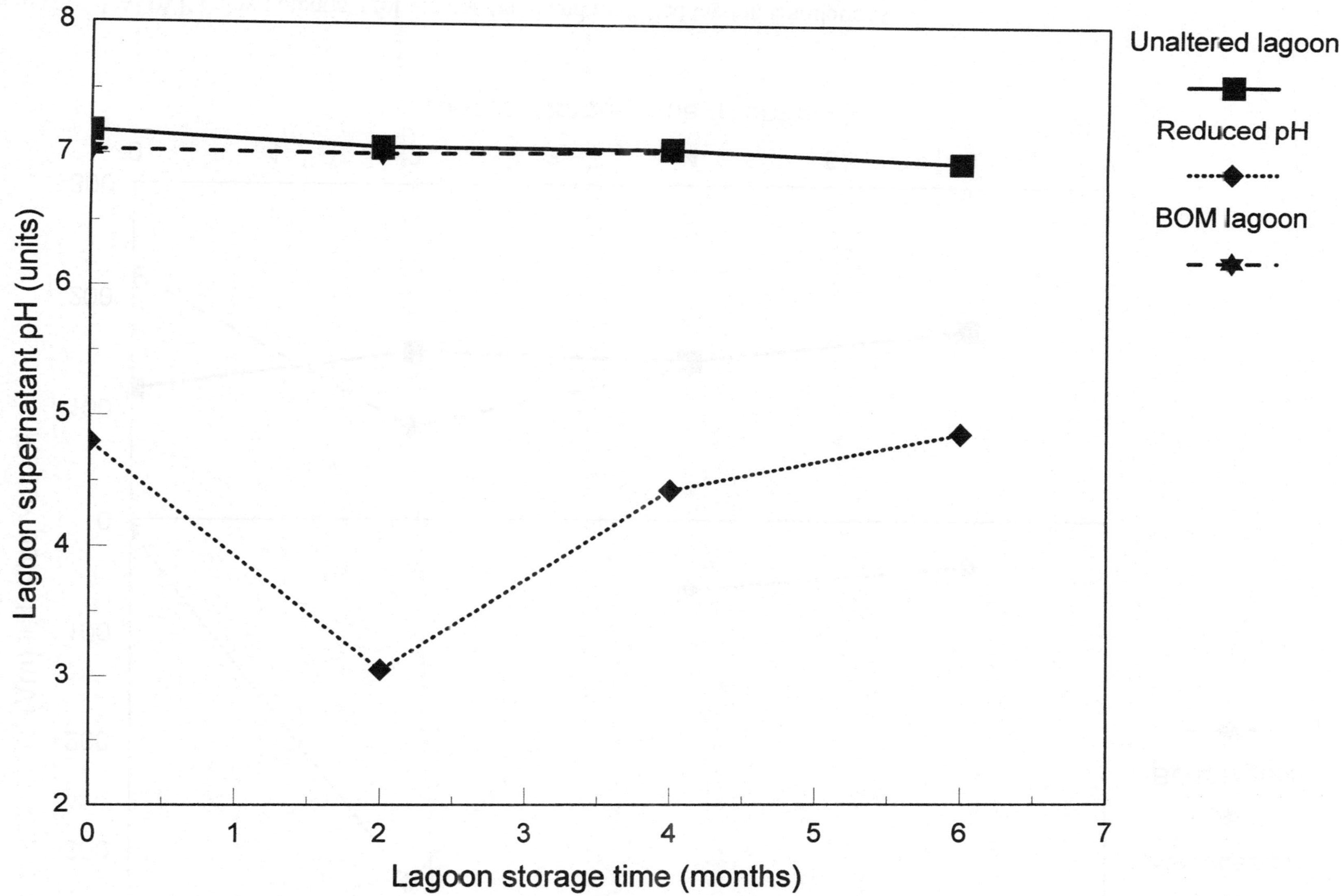

Figure 4.17 LADWP pH data for the unaltered and modified lagoon simulations

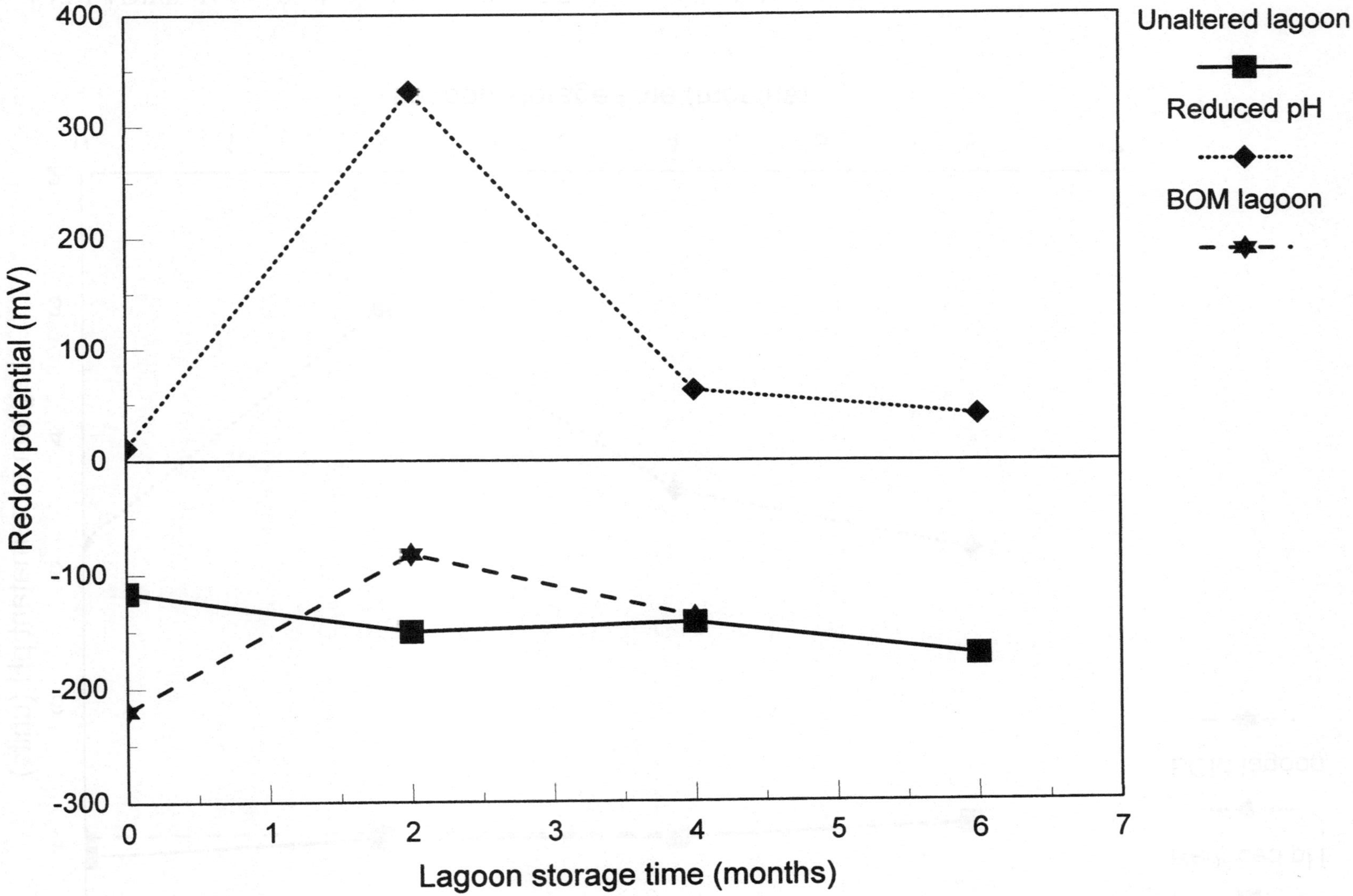

Figure 4.18 LADWP redox potentials for the unaltered and modified lagoon simulations

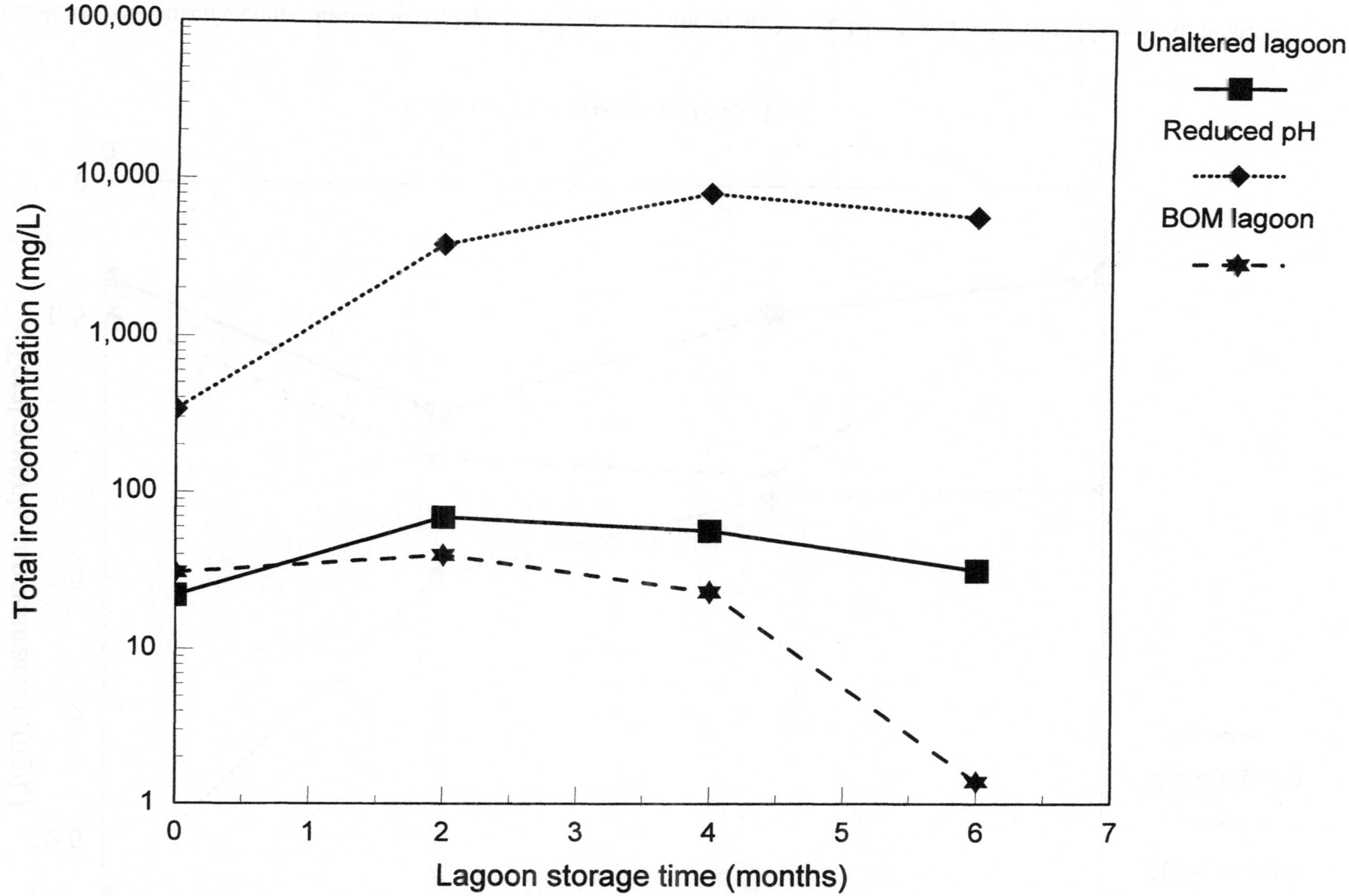

Figure 4.19 Total iron concentration in the lagoon decant over time in unaltered, BOM, and reduced pH lagoon simulations

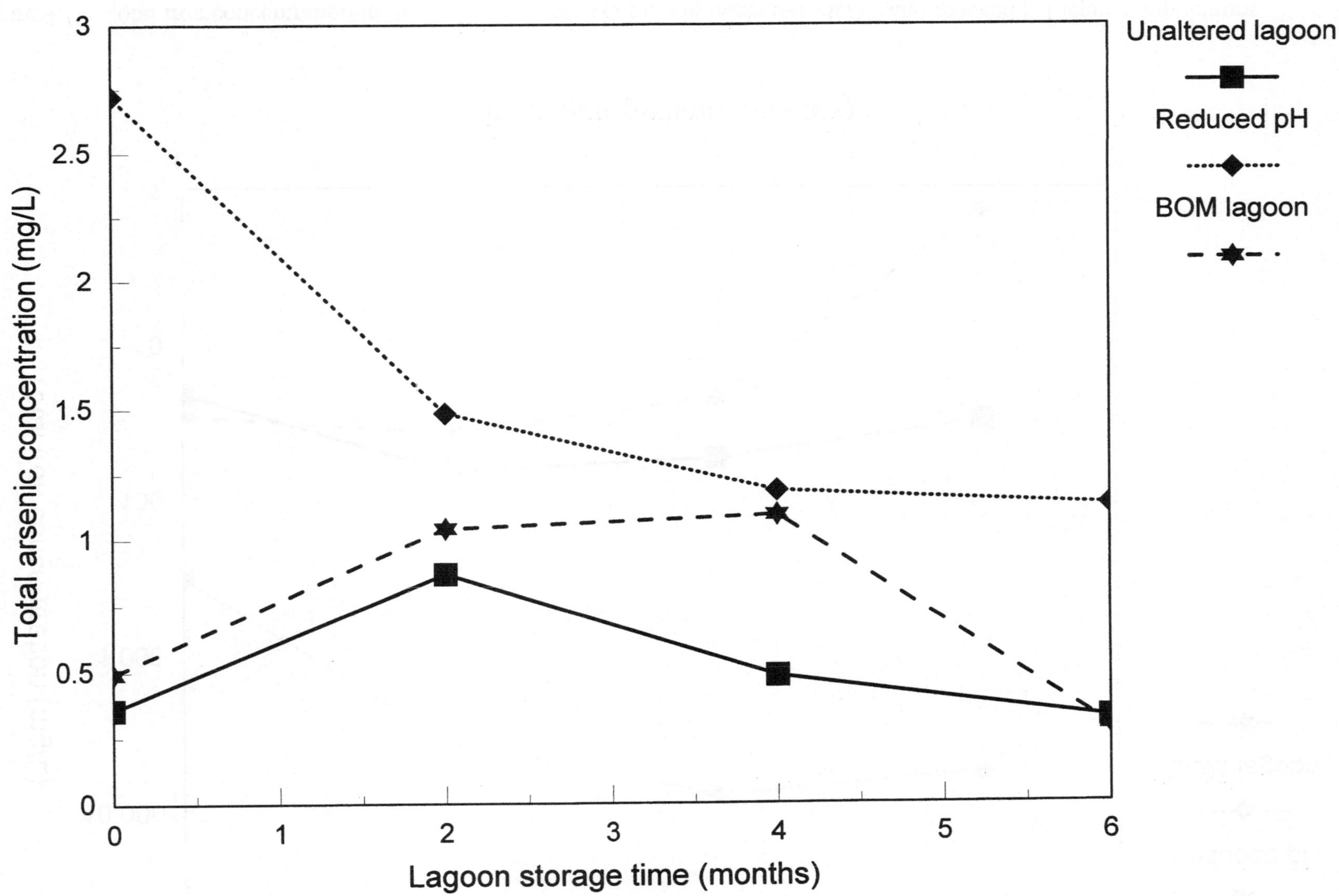

Figure 4.20 Total arsenic concentration in the lagoon decant over time in unaltered, BOM, and reduced pH lagoon simulations

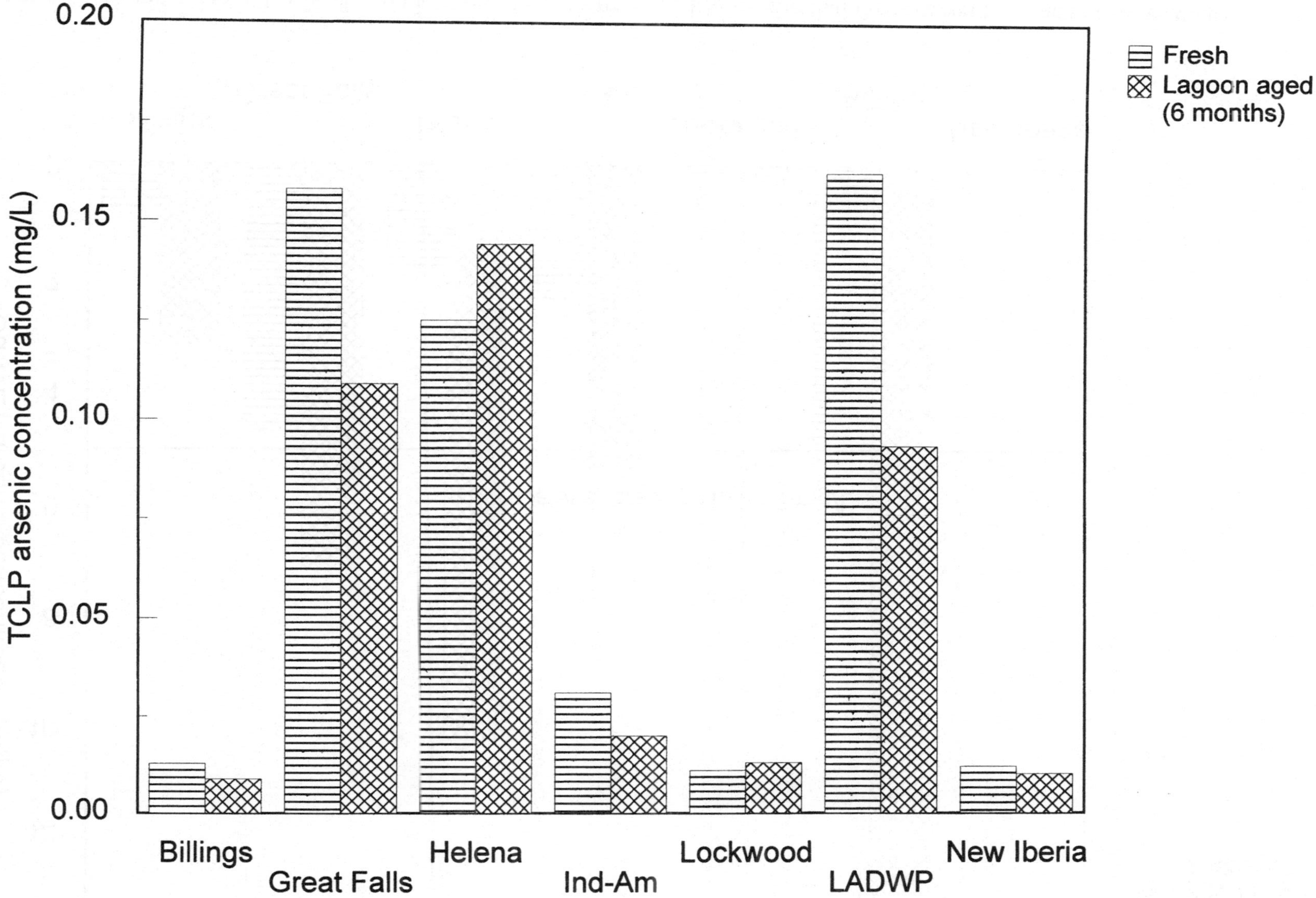

Figure 4.21 Comparison of TCLP arsenic concentrations from the lagoon simulations between fresh and aged residuals

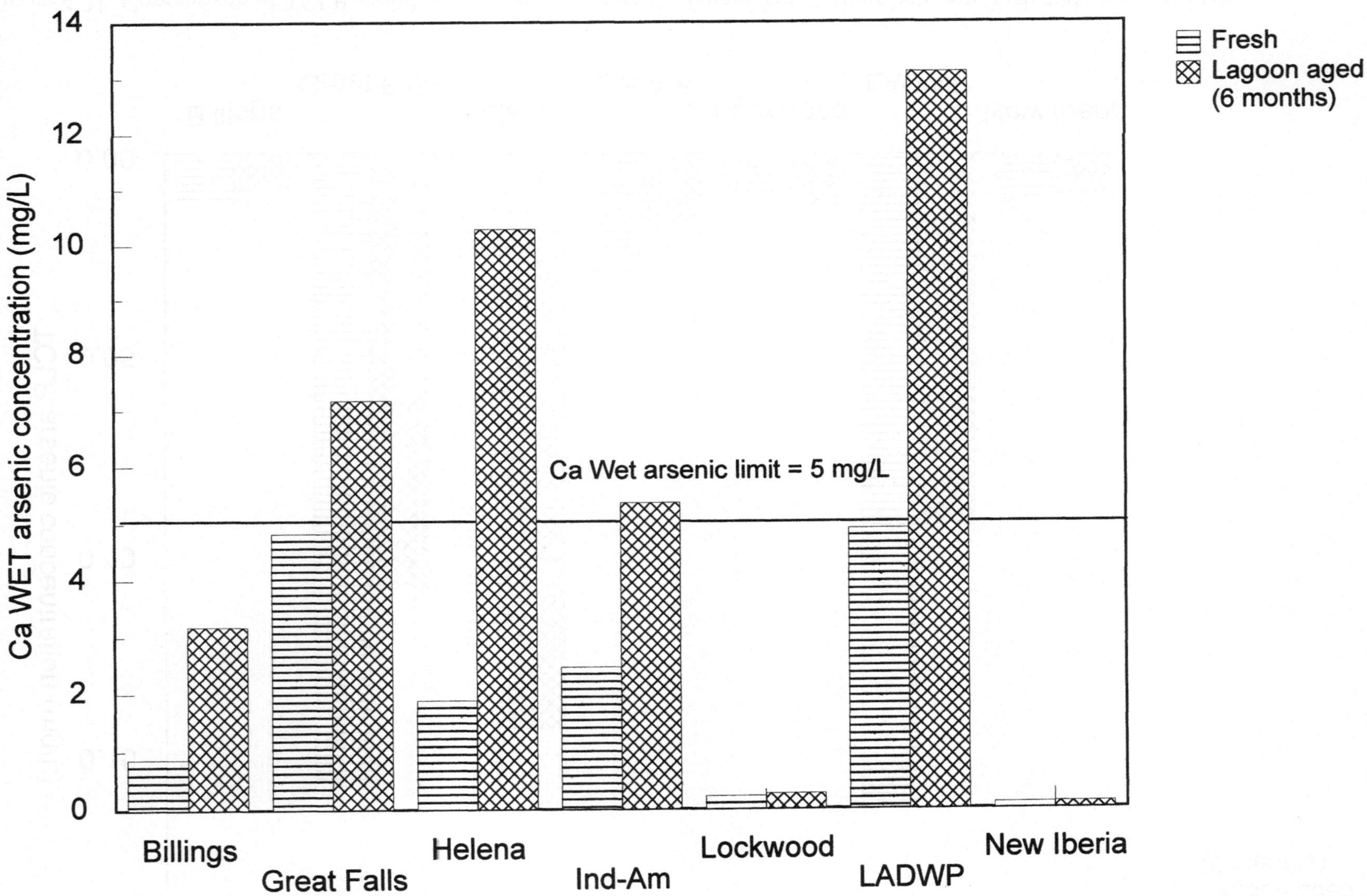

Figure 4.22 Comparison of Ca WET arsenic concentrations from the lagoon simulations between fresh and aged residuals

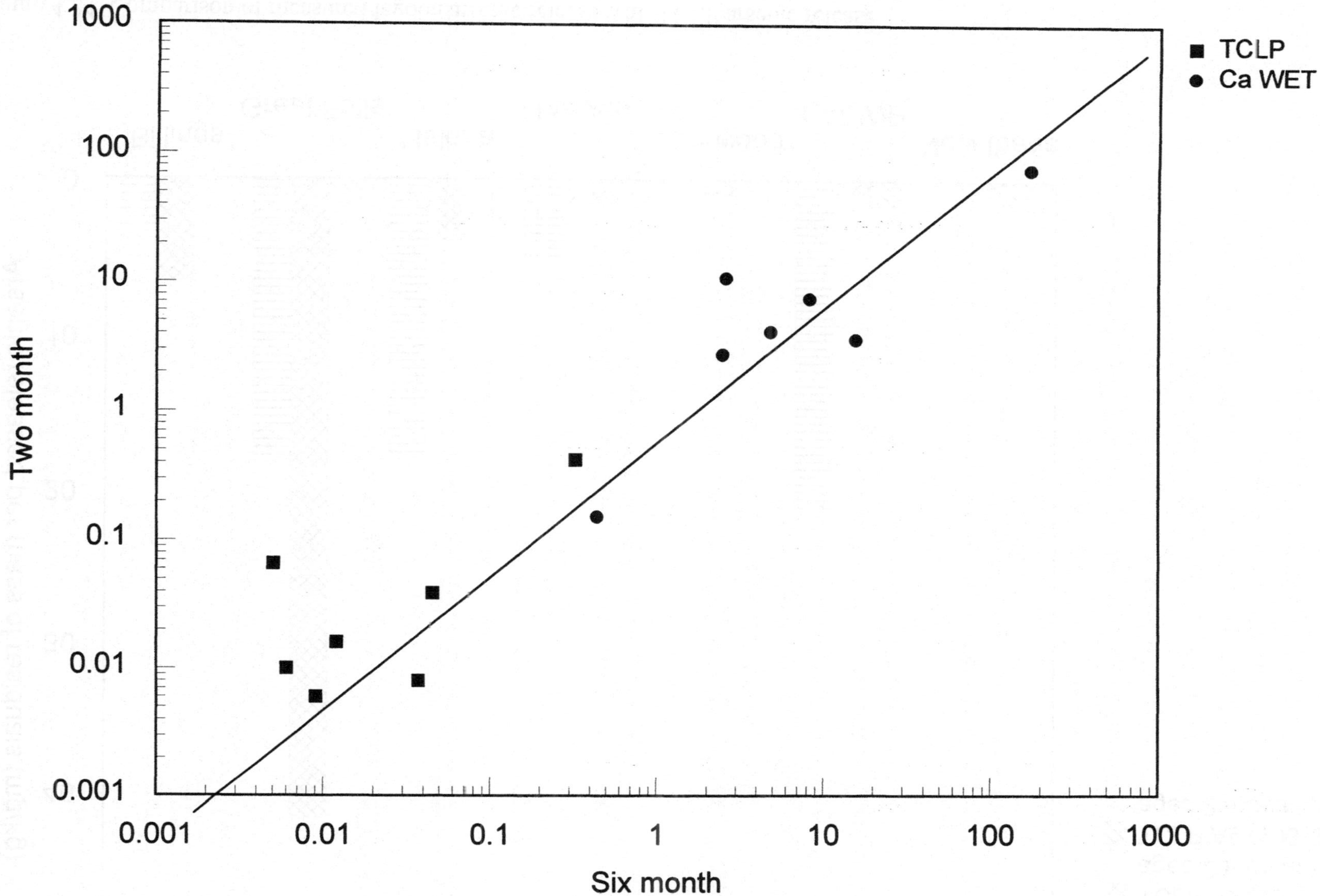

Figure 4.23 Impact of residuals aging on Ca WET and TCLP arsenic levels

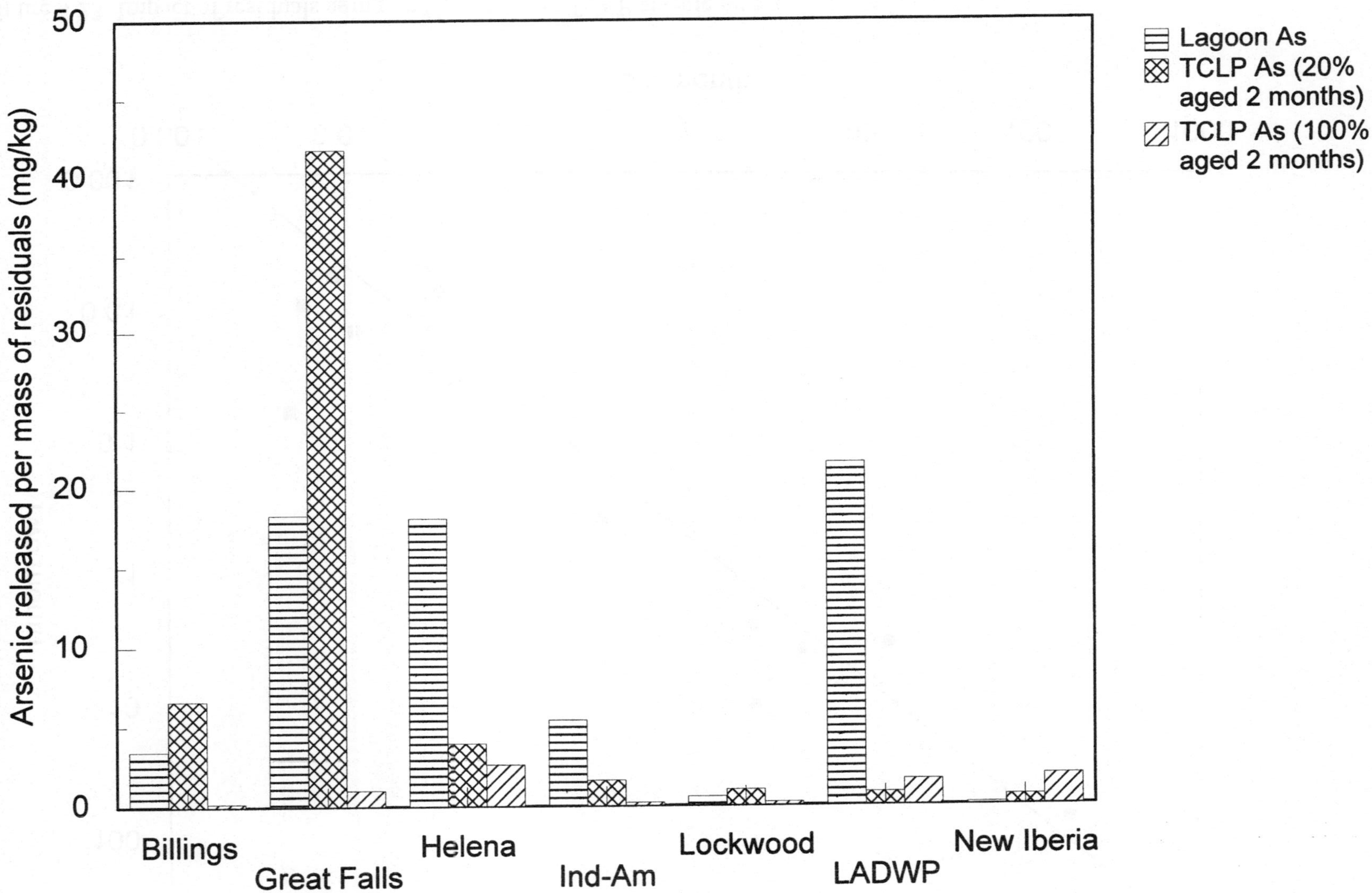

Figure 4.24 Comparison of measured lagoon arsenic release with TCLP arsenic release

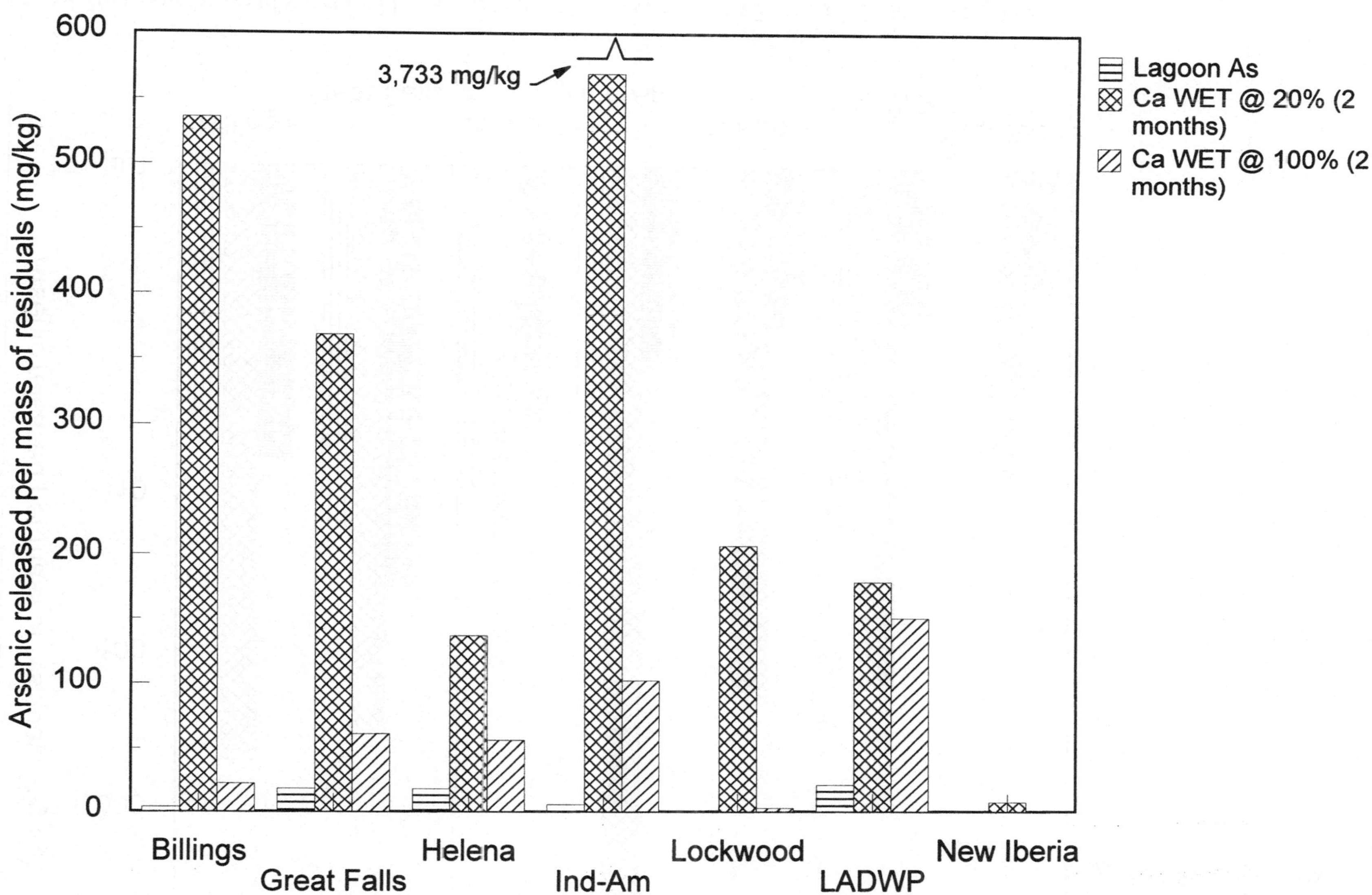

Figure 4.25 Comparison of measured lagoon arsenic release with Ca WET arsenic release

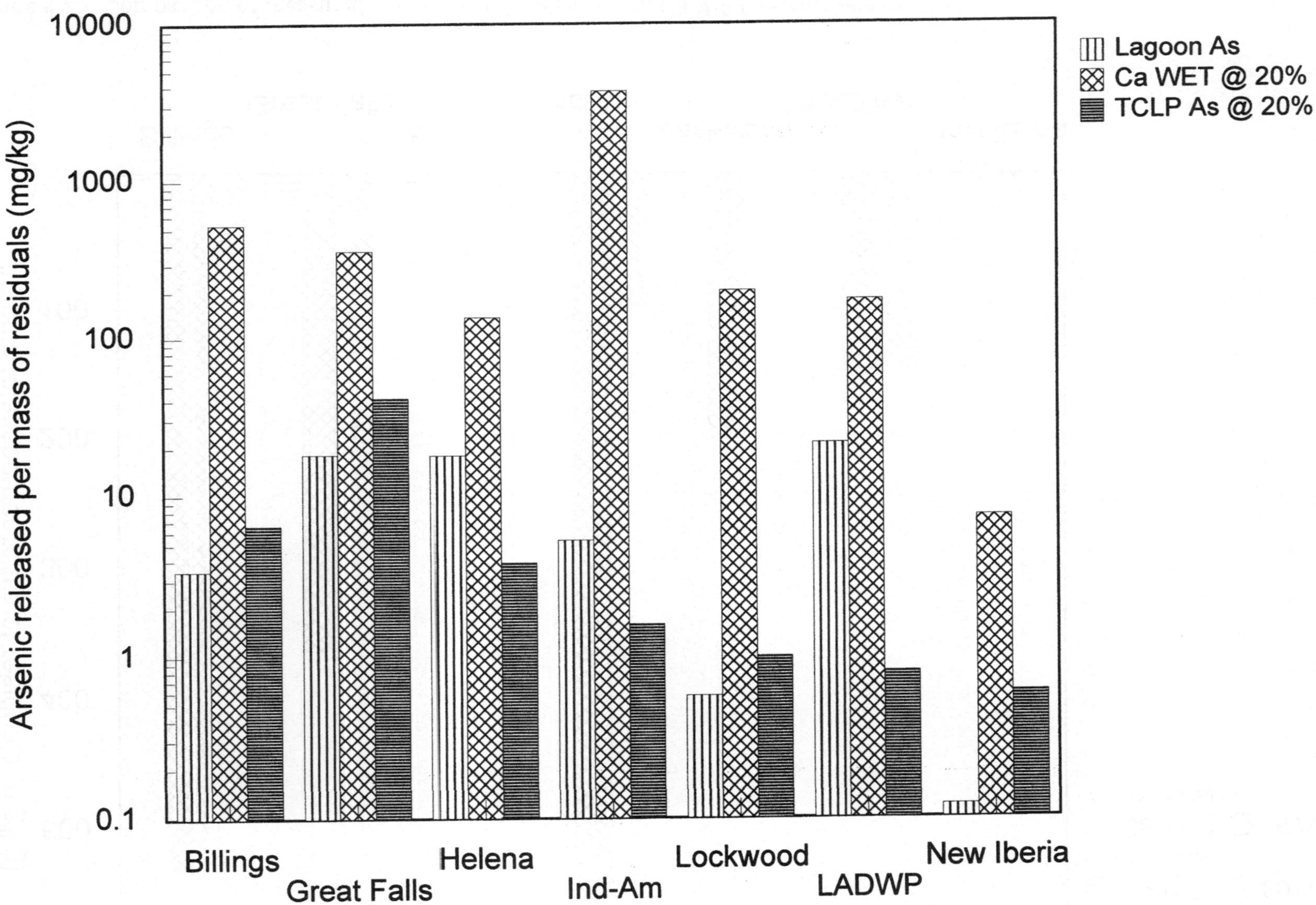

Figure 4.26 Comparison of measured lagoon arsenic release with Ca WET and TCLP at two months

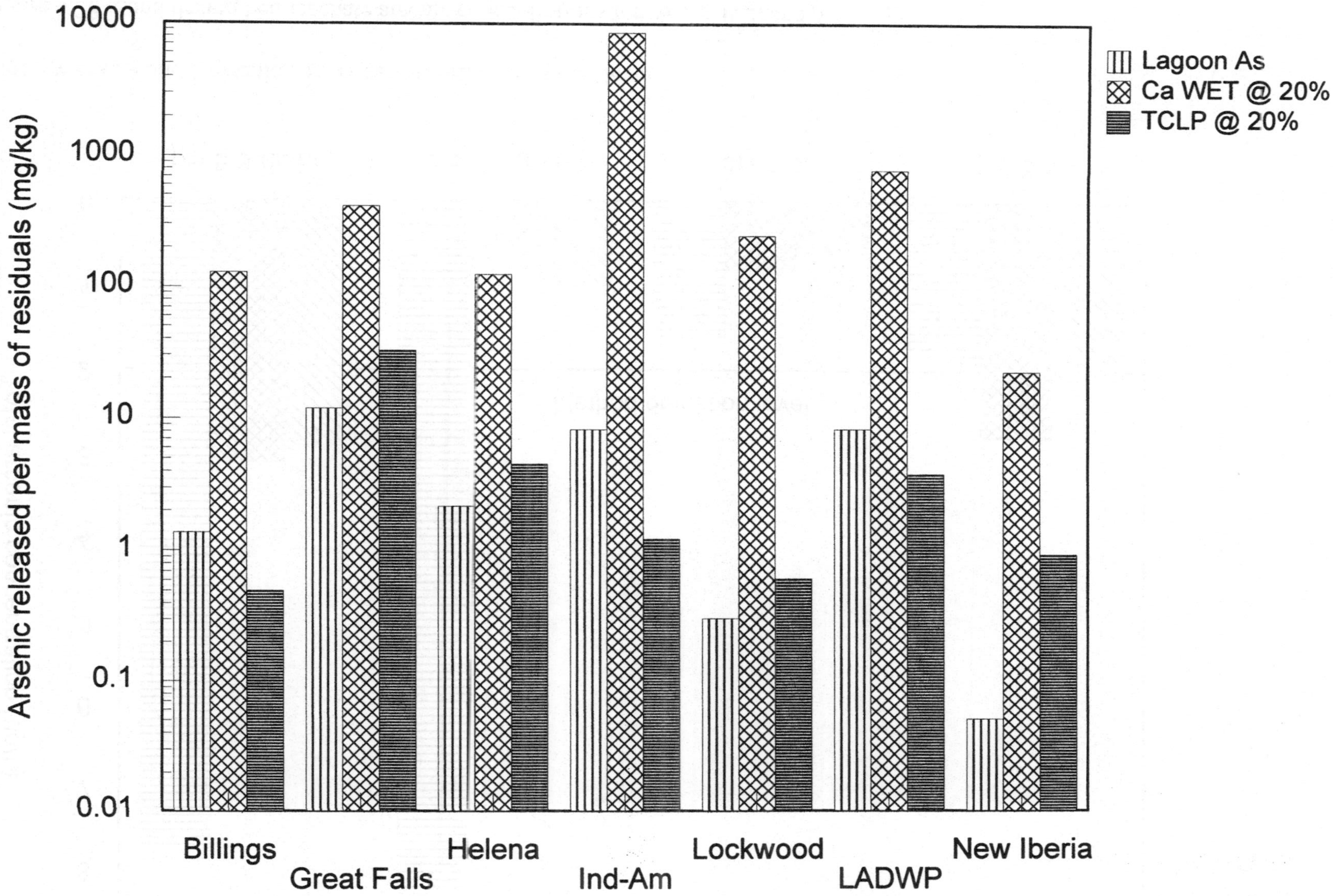

Figure 4.27 Comparison of measured lagoon arsenic release with Ca WET and TCLP at six months

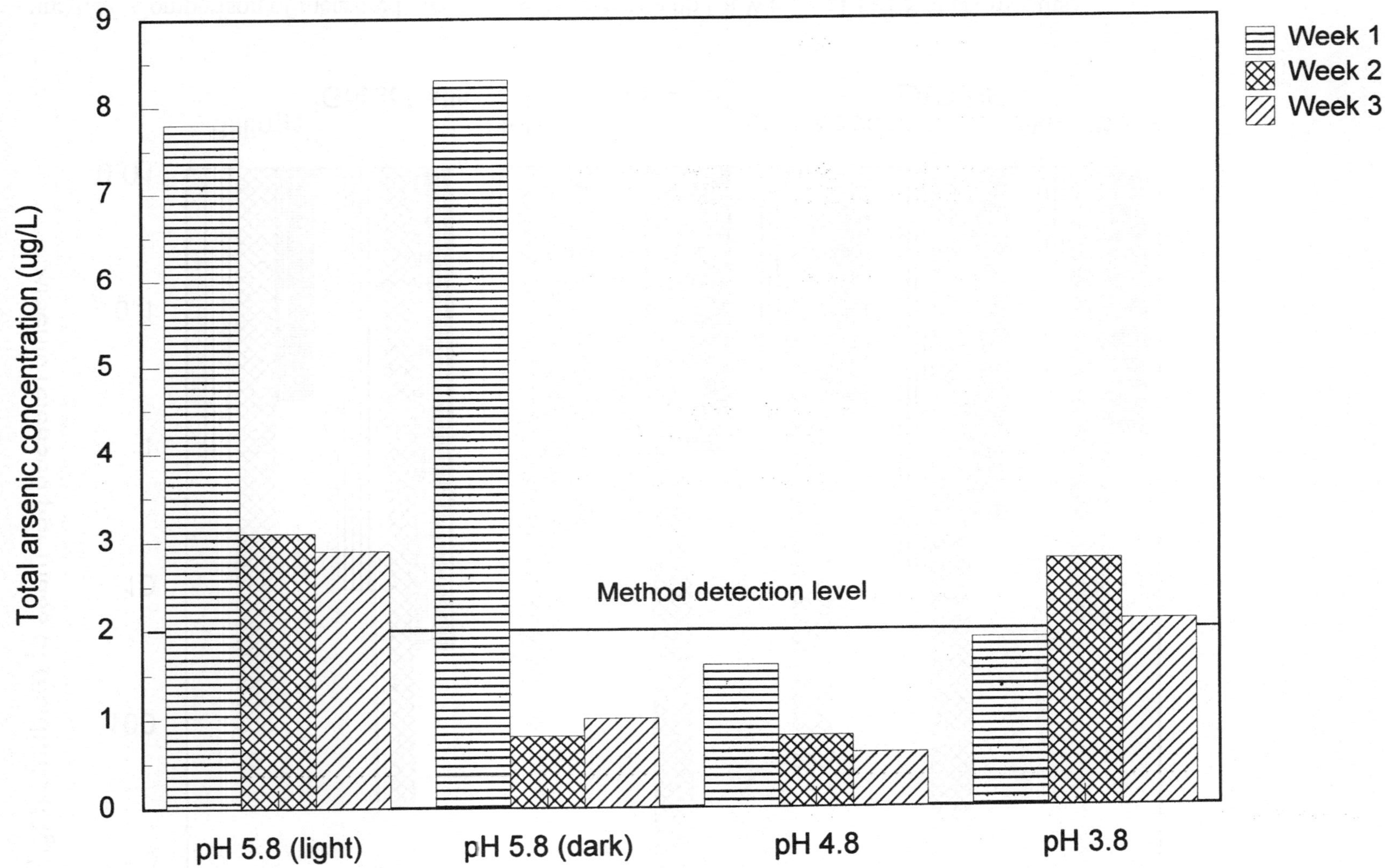

Note: Weeks 4 and 5 arsenic concentration below detection level

Figure 4.28 Sand drying bed leachate arsenic concentrations for weeks 1, 2, and 3

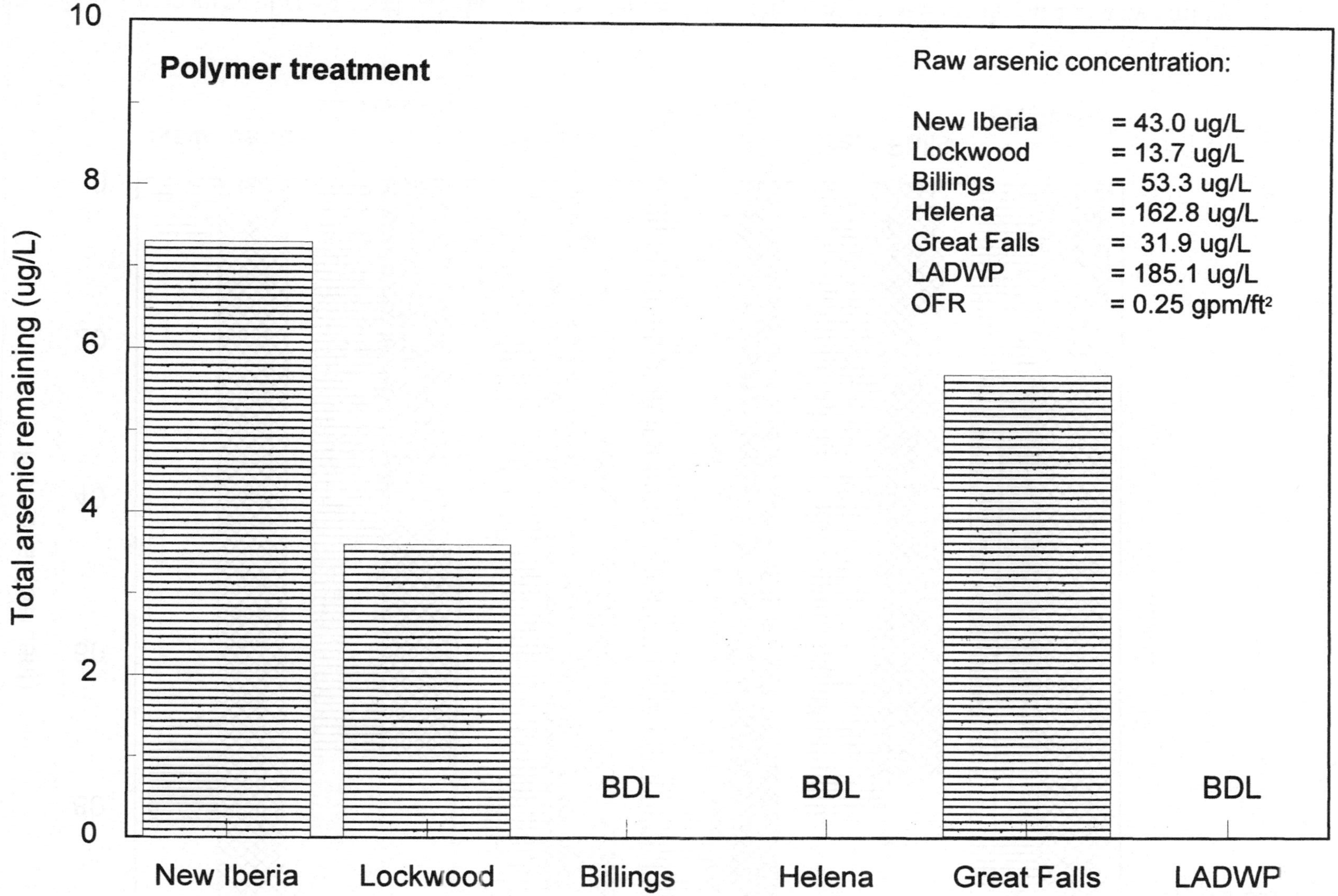

Figure 4.29 Arsenic remaining after settling of SFBW using polymer

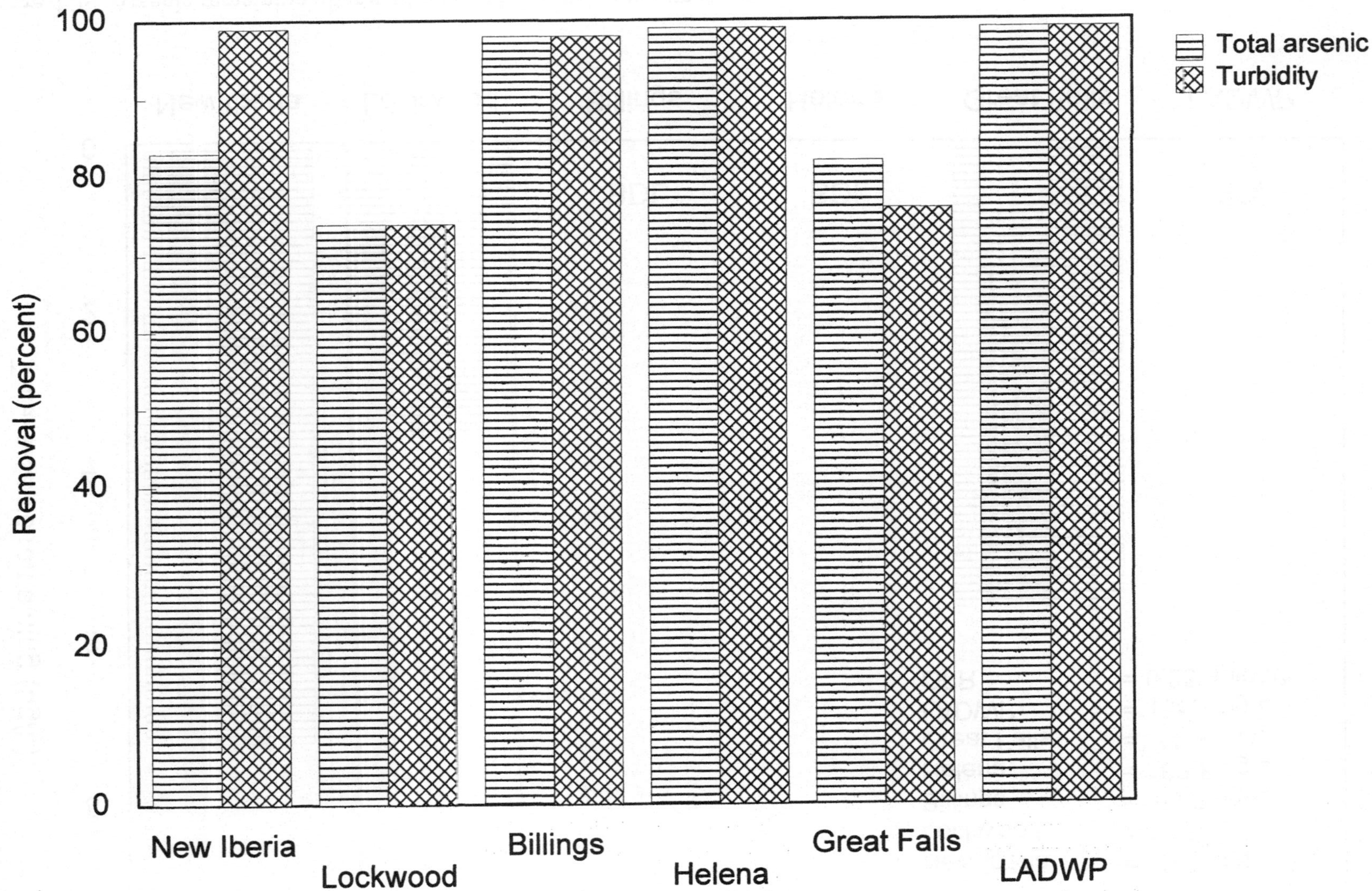

Figure 4.30 Comparison of arsenic and turbidity removal from SFBW using polymer conditioning and gravity settling

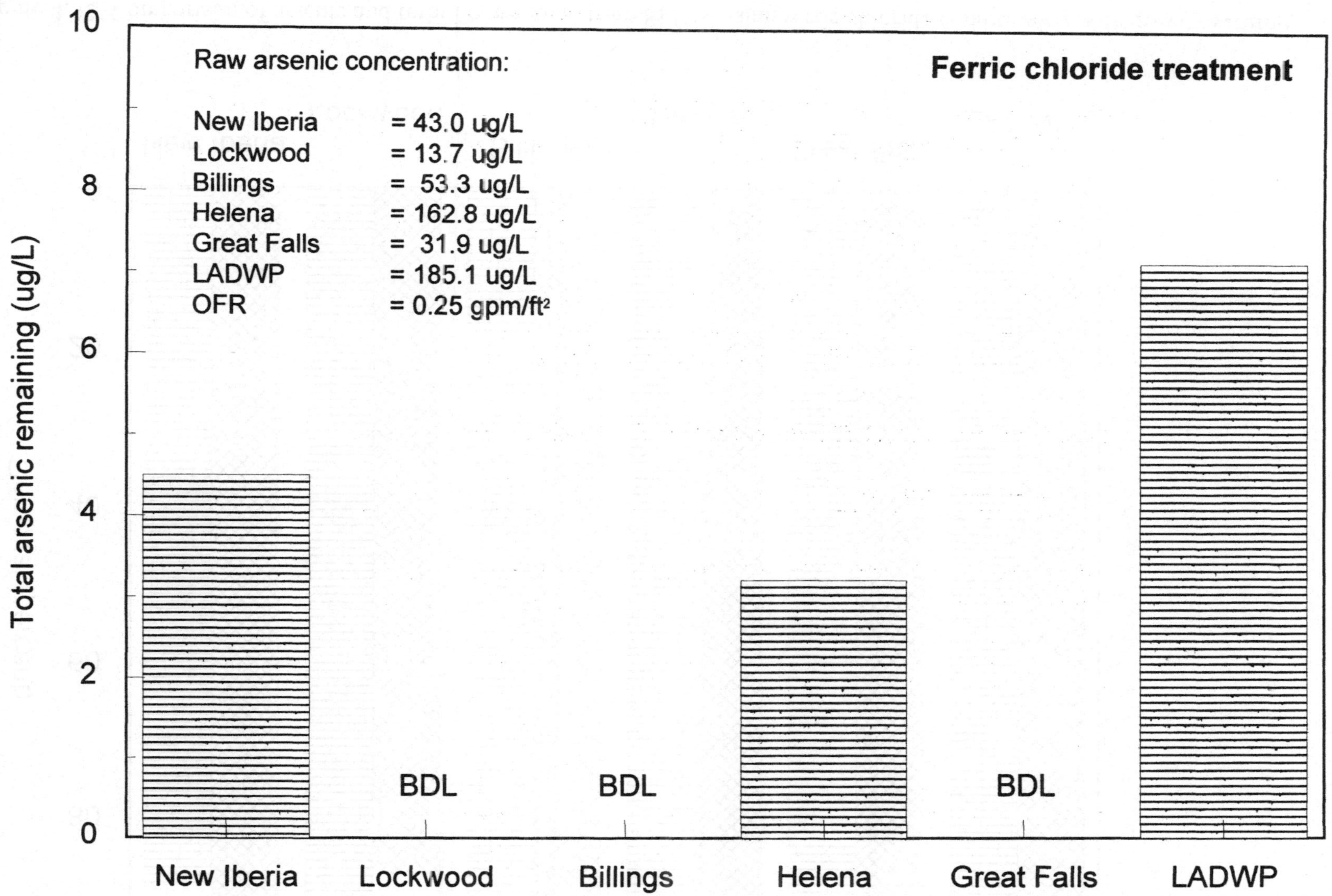

Figure 4.31 Arsenic remaining in solution after treatment of SFBW with ferric chloride

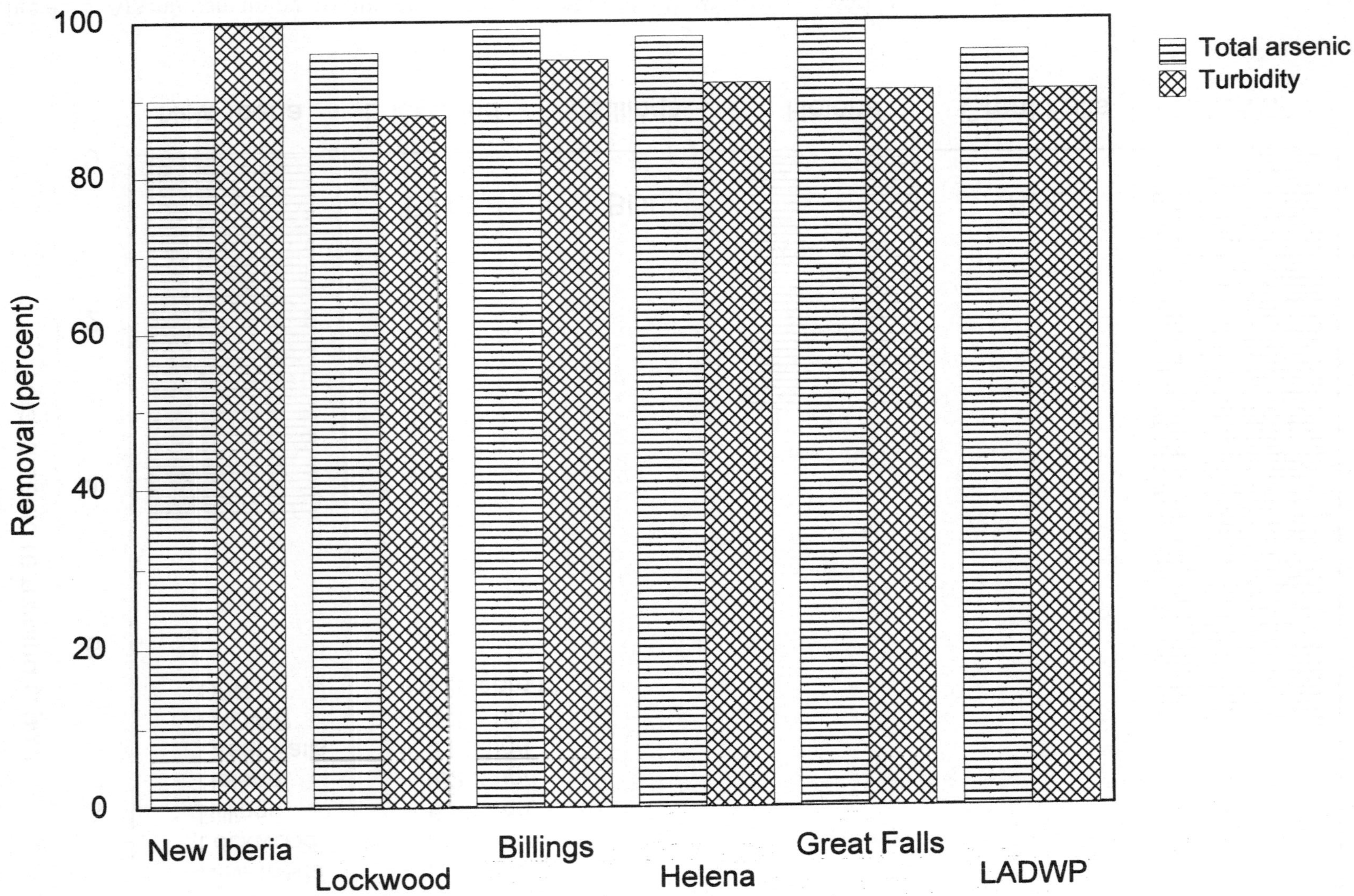

Figure 4.32 Comparison of arsenic and turbidity removal from SFBW using ferric chloride conditioning with gravity settling

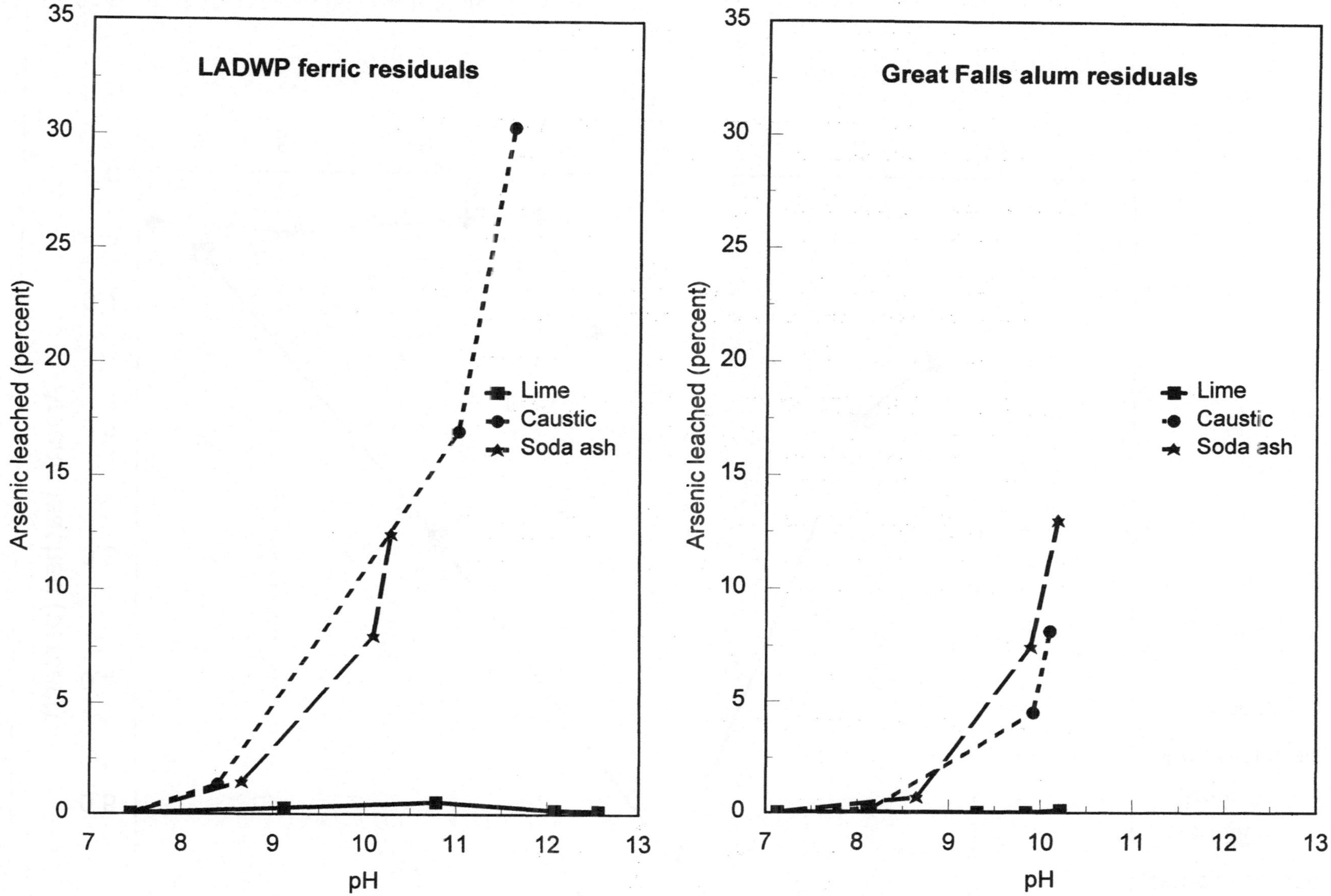

Figure 4.33 Effect of lime, sodium hydroxide, and soda ash on arsenic leachability from ferric and alum residuals

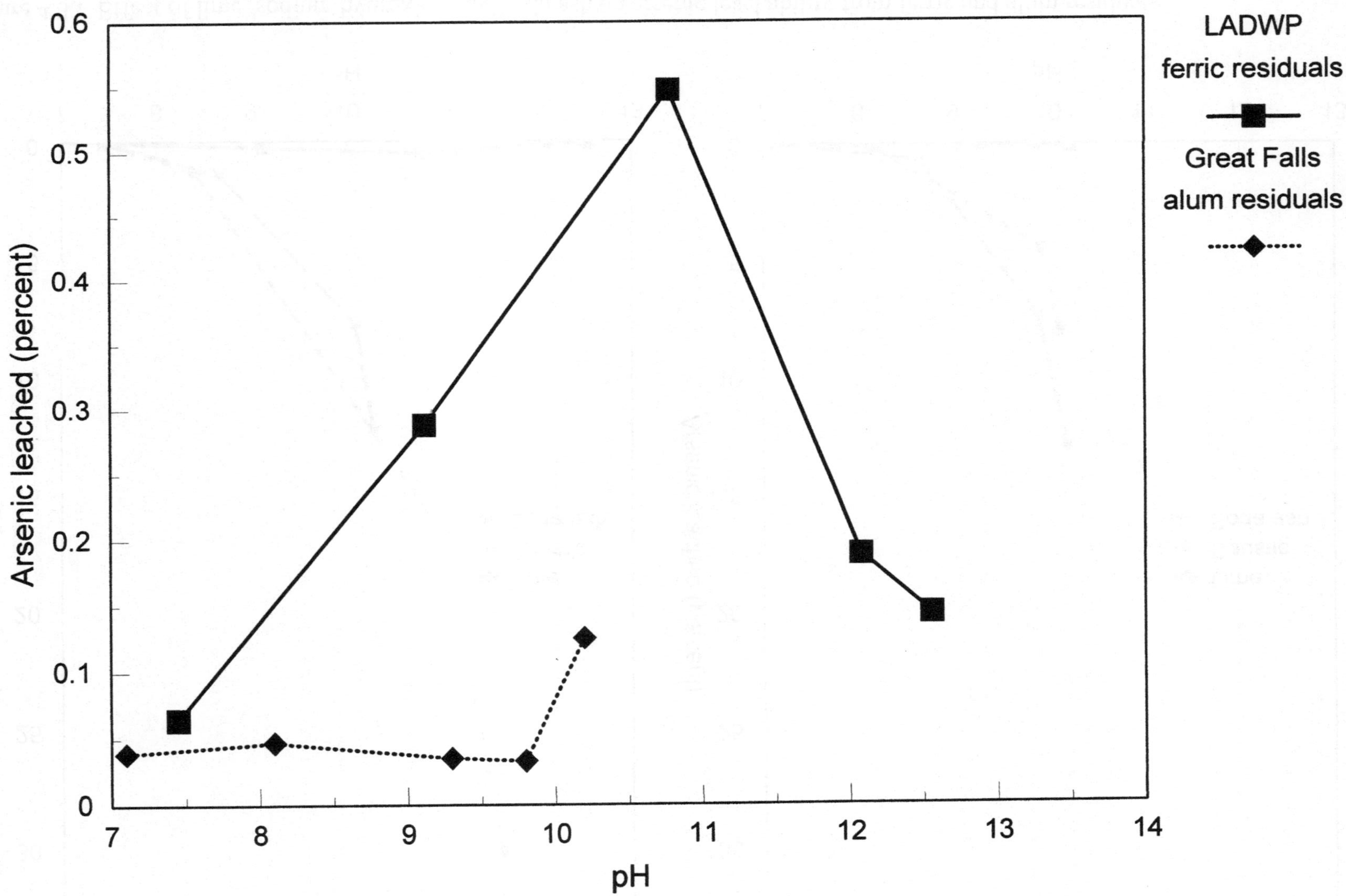

Figure 4.34 Effect of pH on arsenic leachability using lime

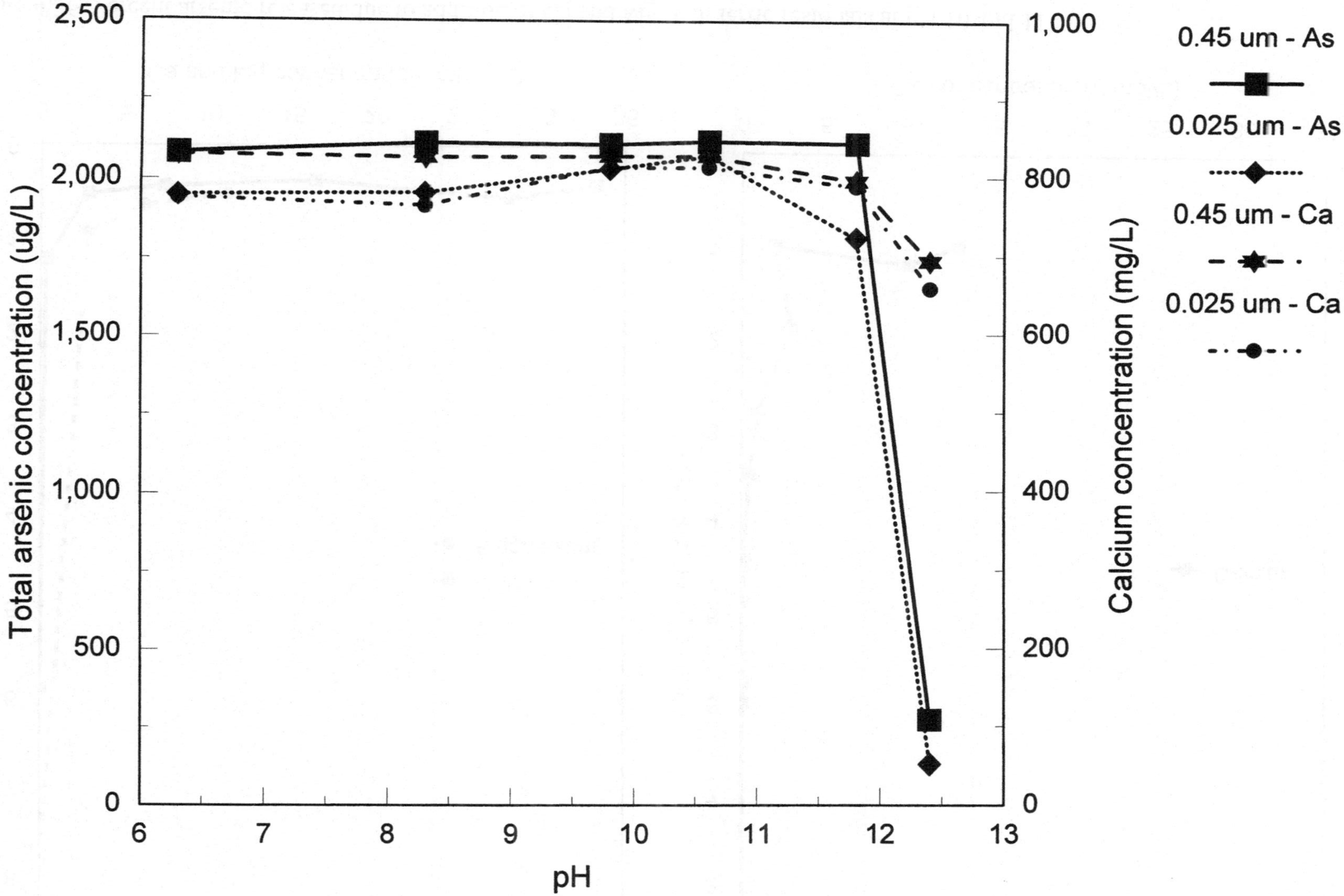

Figure 4.35 pH of formation of calcium arsenate solids

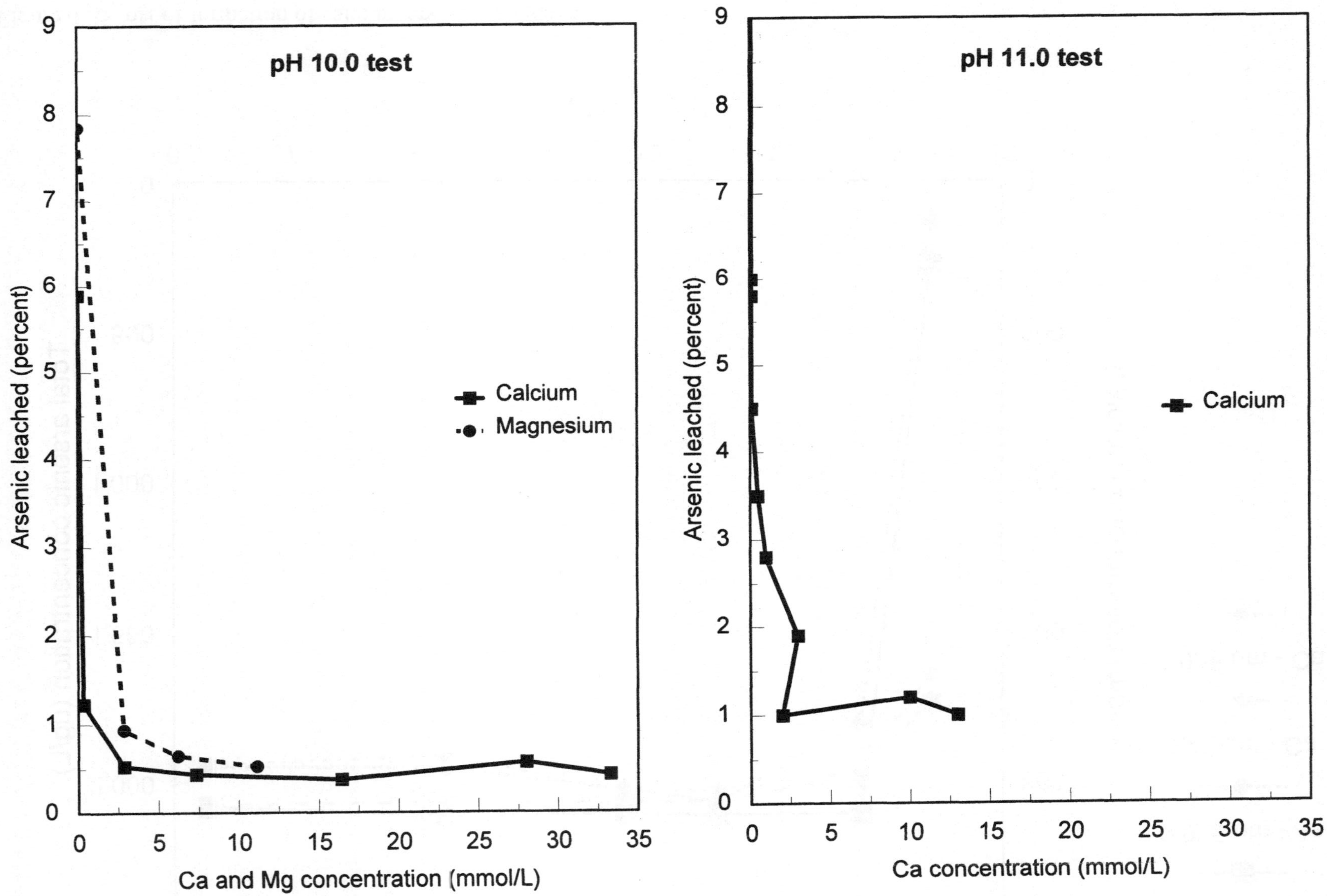

Figure 4.36 Percent arsenic released due to addition of Ca and Mg with ferric residuals at pH 10 and 11

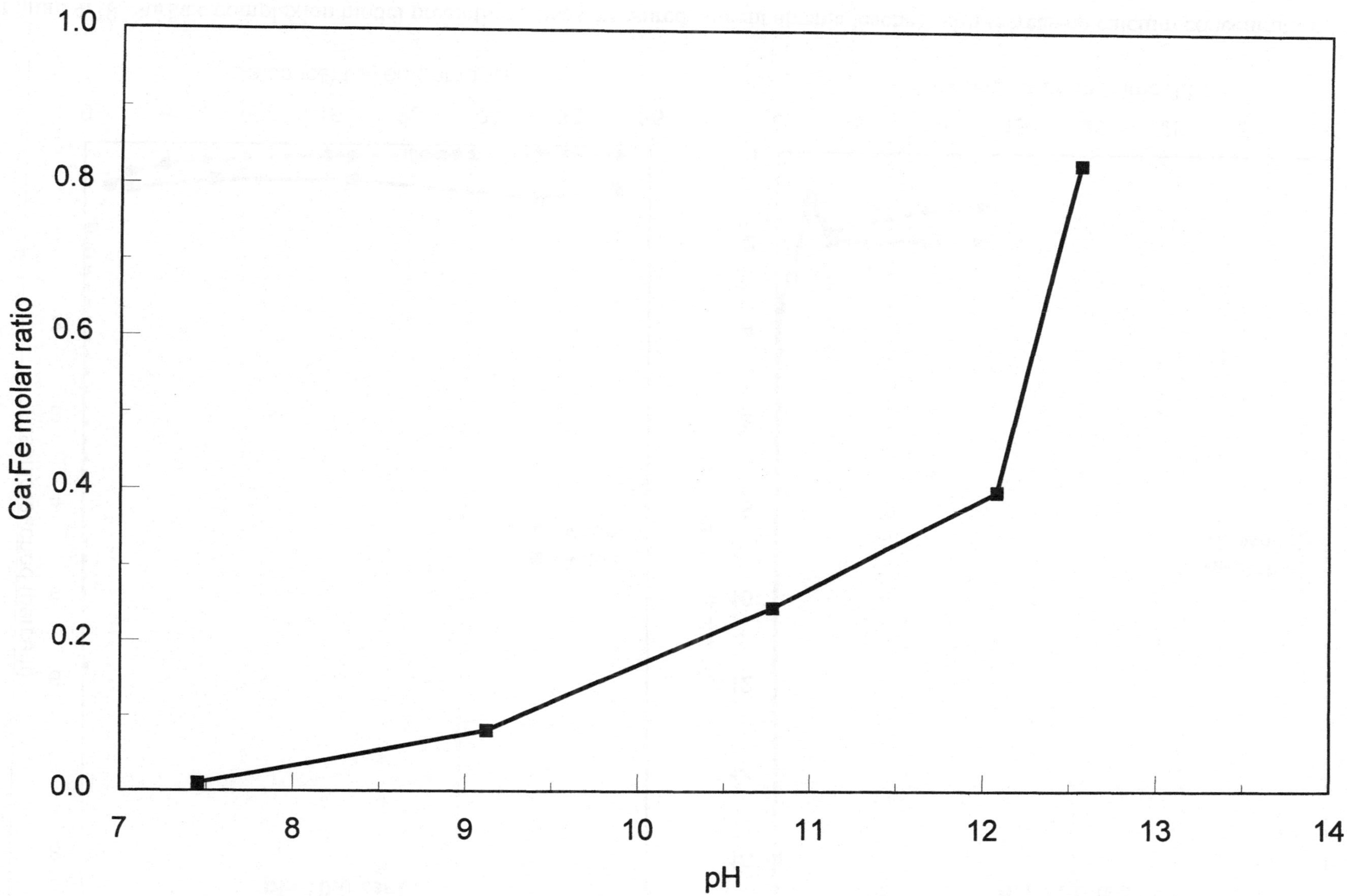

Figure 4.37 Particulate calcium to particulate iron molar ratio versus pH

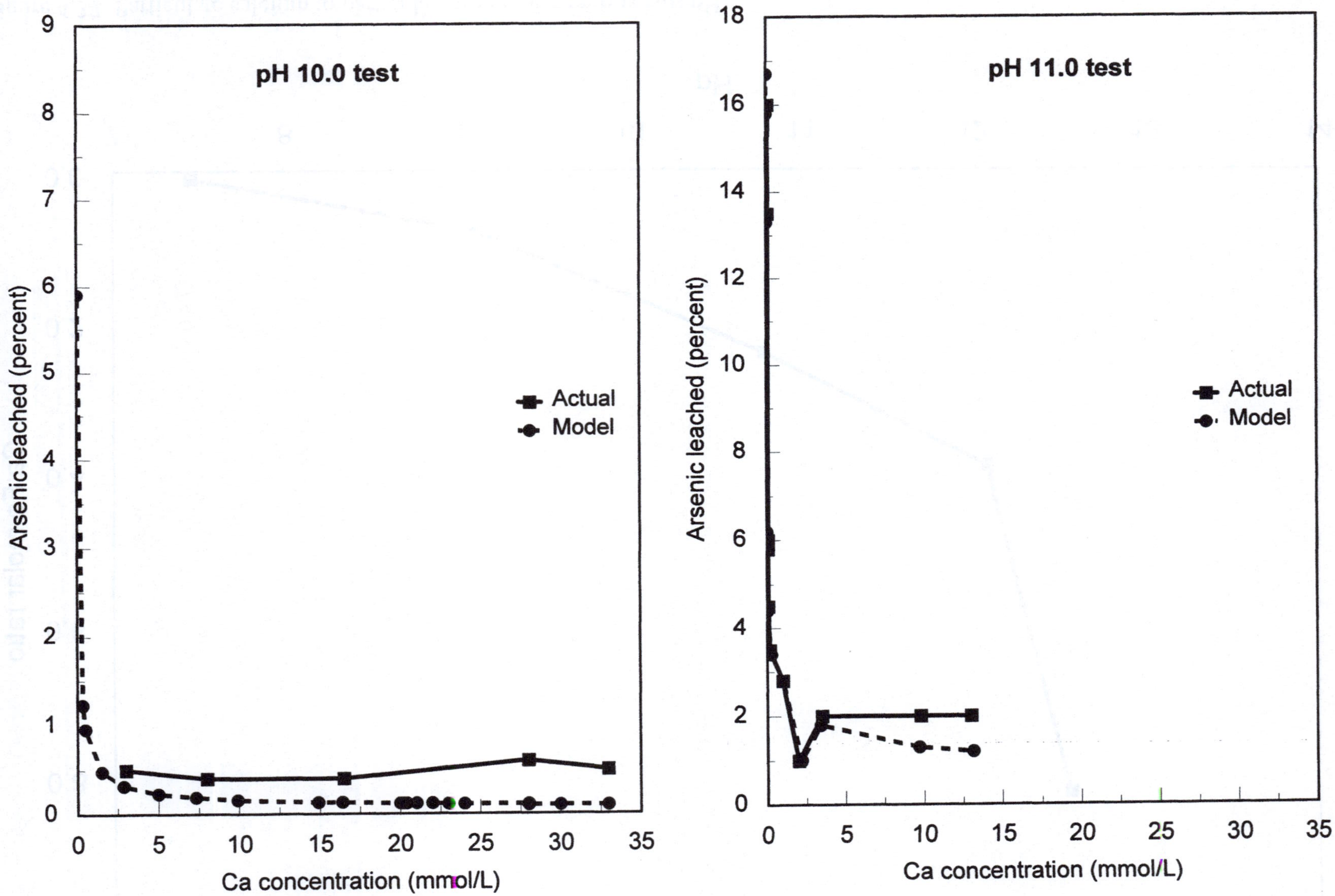

Figure 4.38 Surface complexion model prediction versus measured percent arsenic leached with increasing calcium concentration

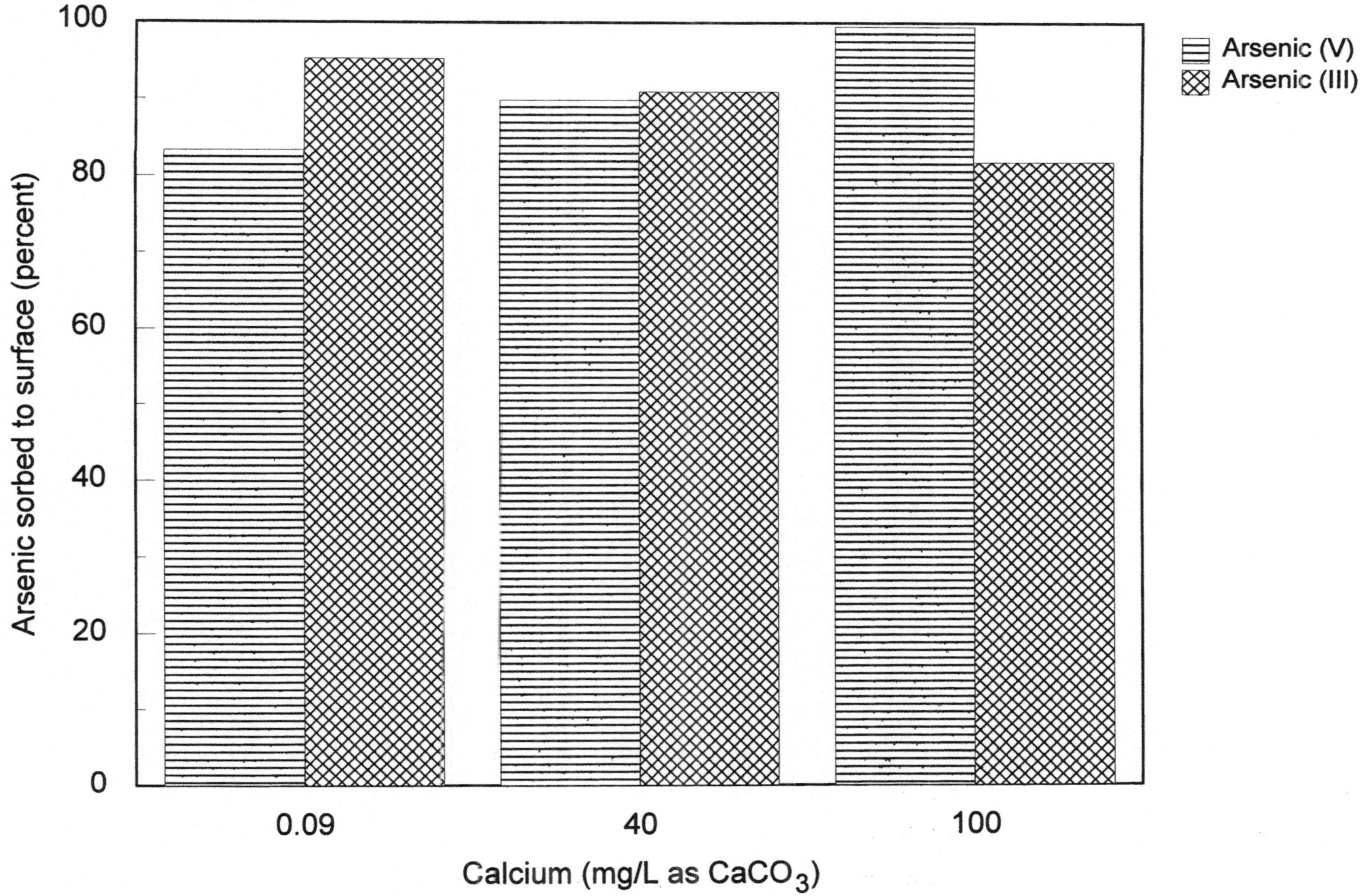

Figure 4.39 Modeling results: Arsenic (V) and arsenic (III) versus calcium at pH 11

CHAPTER 5

UTILITY GUIDELINES FOR RESIDUALS HANDLING

INTRODUCTION

The regulatory maximum contaminant limit (MCL) for arsenic has been lowered to 0.1 mg/L. Reduction of the arsenic MCL will impact utilities in several ways, but one area of increasing concern is the fate of arsenic in residuals. Ultimate disposal options may be limited because a change in the MCL may result in residuals being classified as hazardous wastes and thus no longer acceptable for disposal by municipal landfills. Land application of WTP residuals is already limited by arsenic levels in some areas, and additional removal requirements could impact this further. Indirect discharge of residuals may also be restricted because the arsenic will ultimately impact the acceptability of wastewater biosolids for land application. Liquid residuals resulting from solid/liquid separation (overflows from lagooning, dewatering, filter backwash water recovery) may no longer be acceptable for sewer or stream discharge. These side streams might require special treatment processes, or the entire residuals handling process could require modification.

If the residuals do not pass the TCLP test, or the WET test in California, the residuals will be classified as a hazardous waste and expensive handling and disposal techniques will be required unless fixation methods that could bind the arsenic more tightly to the solids. For utilities, the goal of this research was to provide a guide that could be used to assist in determining the best choice in technologies for arsenic removal after considering the whole treatment train, including residuals treatment requirements, final disposal options, as well as overall costs. These factors will be expected to differ for various treatment techniques, different chemicals used in treatment, and for varying concentrations of arsenic in the raw water and residuals. This chapter is intended to combine the findings obtained from this research effort, along with corresponding research work performed by the authors for a recent EPA project (USEPA 2001) and other related arsenic projects. The objective of this guideline section is to synthesize the characteristics, leaching properties, treatability options, disposal options, and costs for handling and disposal of arsenic-laden residuals into a document that can be readily used by utilities in their decision-making process. This document should assist utilities that need to either select an arsenic removal treatment process or evaluate required modifications for an existing treatment system.

RESIDUALS DISPOSAL PATHWAYS

There are a number of treatment techniques available for removal of arsenic from municipal drinking water. Each of these processes would generate some form of residual that would contain high concentrations of arsenic or other contaminants removed from raw water. The processes that are most frequently used for arsenic removal applications are listed in Table 5.1.

Table 5.1

List of pretreatment processes used for arsenic removal

	Residuals generated	
Treatment process	Liquids	Solids
Activated alumina (no media regeneration)	SFBW	AA media
Activated alumina (w/media regeneration)	AA regenerant	AA media
Ion exchange	IX regenerant	IX media
Manganese greensand filtration*	SFBW	Filter media
Fe-based media adsorption	SFBW	Fe-based media
Coagulation-microfiltration (CMF with no clarification)†	Membrane backwash	None‡
Conventional treatment (clarification, granular media filtration)	SFBW, clarifier blowdown	Settled solids
Lime softening w/granular media filter	SFBW, clarifier blowdown	Settled solids
Reverse osmosis (RO)	Concentrate	None
Nanofiltration (Nf)	Concentrate	None

*And other Fe/Mn removal processes including air oxidation/chemical oxidation

†If a clarifier is used in the CMF process, clarifier blowdown would be a second liquid residuals stream

‡Solids would be generated by dewatering of backwash residuals

For utilities that are in the process of selecting an arsenic removal treatment technology, it is critical to evaluate the types of residuals that would be generated, their expected arsenic concentrations, and the residuals pre-treatment strategies that would be required prior to final disposal. A utility needs to understand the residuals impacts from removing arsenic in the raw water and, specifically, if these residuals will be able to meet: (1) NPDES as limits (for direct discharge), (2) POTW limits (indirect discharge), (3) land application limits, and (4) the TCLP arsenic limit for non-hazardous landfill disposal (with or without pre-treatment before final disposal).

To demonstrate the possible pathways for residuals handling and disposal, a flowchart was developed and is included in Figure 5.1. The flow chart shows the expected residuals handling scenarios for residuals that contain sufficiently high arsenic concentrations such that direct or indirect disposal is not possible. The residuals handling pathways for each of the treatment processes listed in Figure 5.1 are described in the following summaries.

Activated Alumina

Activated alumina systems could operate using either media replacement after exhaustion or on-site media regeneration. For most systems, replacement of the media may be a more economical and practical approach to construction and operations of facilities for treatment of the regenerate stream. When AA with no regeneration is utilized, the solid residual generated is the spent AA media that can likely be disposed of in a non-hazardous waste landfill. Ability of spent AA media to pass the California WET test is unknown. These media will still often require backwashing that results in SFBW that may require facilities treatment prior to discharge.

When regeneration of AA is used, a liquid residual is generated for disposal along with eventually the spent media. There are two different options to handling the liquid AA regenerate:

- Treatment (could include precipitation, adsorption, or fixation), or
- Evaporation lagoon

Treatment of AA liquid residuals could include a number of different technologies including chemical precipitation using ferric chloride (with or without polymer) and Fe-based media adsorption. Figure 5.1 demonstrates the basic flow pathway for the precipitation process that is used to remove suspended or dissolved arsenic from the liquid residuals and concentrate it in solid form. Thus, generation of a more concentrated solid will occur.

If treatment is used, then the goal is to reduce arsenic levels such that the liquid stream can be disposed of by direct or indirect discharge or possibly recycled. Based on data collected during the recent EPA project (USEPA 2001), the solids would pass the TCLP test for arsenic and could be disposed of in a non-hazardous waste landfill. Due to high arsenic levels in the solid residuals, land application may not be possible.

If a utility elects not to remove arsenic from the liquid residuals, the only other option would be to dewater in an evaporation lagoon. The solid residuals from the lagoon after drying could then be landfilled. It should be noted that evaporation lagoons would likely only be practical for utilities that generate small volumes of liquid residuals due to the large land footprint required for drying. In the schematic it is assumed that only the AA, IX, RO, and NF processes could potentially use the evaporation lagoon process due to the typically low volumes of residuals generated and lower concentrations of suspended solids present.

Ion Exchange

The ion exchange (IX) process will generate a liquid residual from the regeneration process. The IX regenerant would need to be handled similarly to the AA regenerate using the same types of treatment options or by using an evaporation lagoon.

Manganese Greensand Filtration and Iron-based Media Adsorption

Although these two treatment processes are different, the type of residuals generated would be very similar. These residuals would include SFBW and spent filter media. The media for each could be landfilled, while the SFBW residuals would require treatment prior to final disposal.

Coagulation Microfiltration, Conventional Treatment, Lime Softening

Coagulation-microfiltration (CMF) with no clarification, conventional treatment using granular media filtration, and lime softening systems would each generate a liquid residual for disposal. The CMF process would generate a backwash stream, while conventional and lime softening would generate SFBW and settled solids blowdown from clarification. Although split treatment is possible for SFBW and clarifier blowdown, the schematic shows that the residuals are combined for treatment. In this research arsenic concentrations were dramatically lowered by separating the solids from the liquid residuals. This reduction is due to the fact that much of the arsenic is primarily bound to iron or other suspended particulate matter in solution.

Summary of Treatment and Disposal Options

The residuals handling diagram is presented in Figure 5.1. The figure summarizes the findings for residuals treatment and disposal from both (USEPA 2001) and this research. The figure lists the suitable disposal pathways for the residuals resulting from the various arsenic removal treatment processes along with potential treatment options.

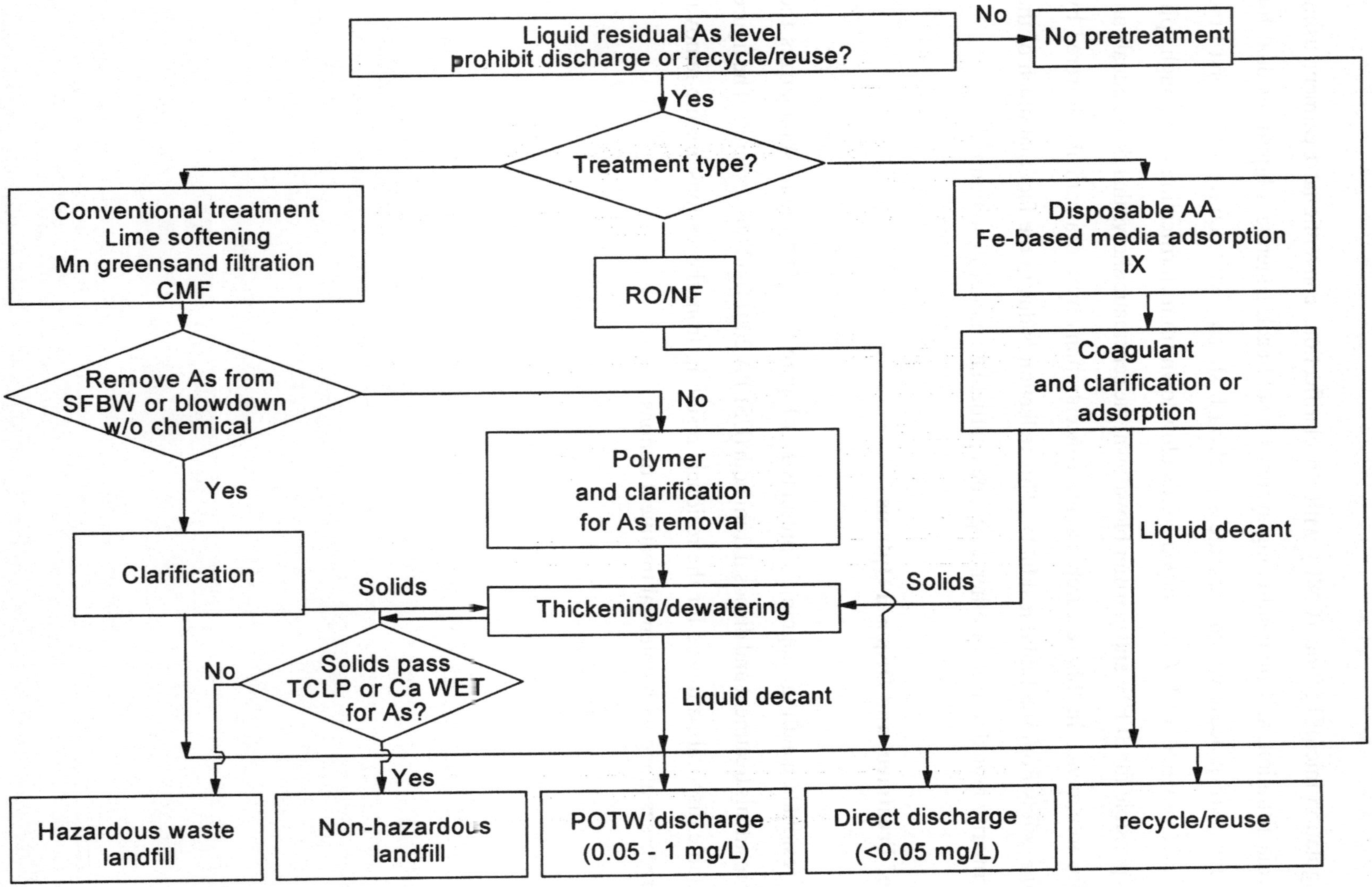

Note: Spent media disposed of in non-hazardous waste landfill

*AA and iron-based media adsorption backwash waters expected to meet POTW direct discharge or recycle arsenic criteria

Figure 5.1 Arsenic residuals handling and disposal decision tree

APPENDIX A

PLANT FLOW SCHEMATICS

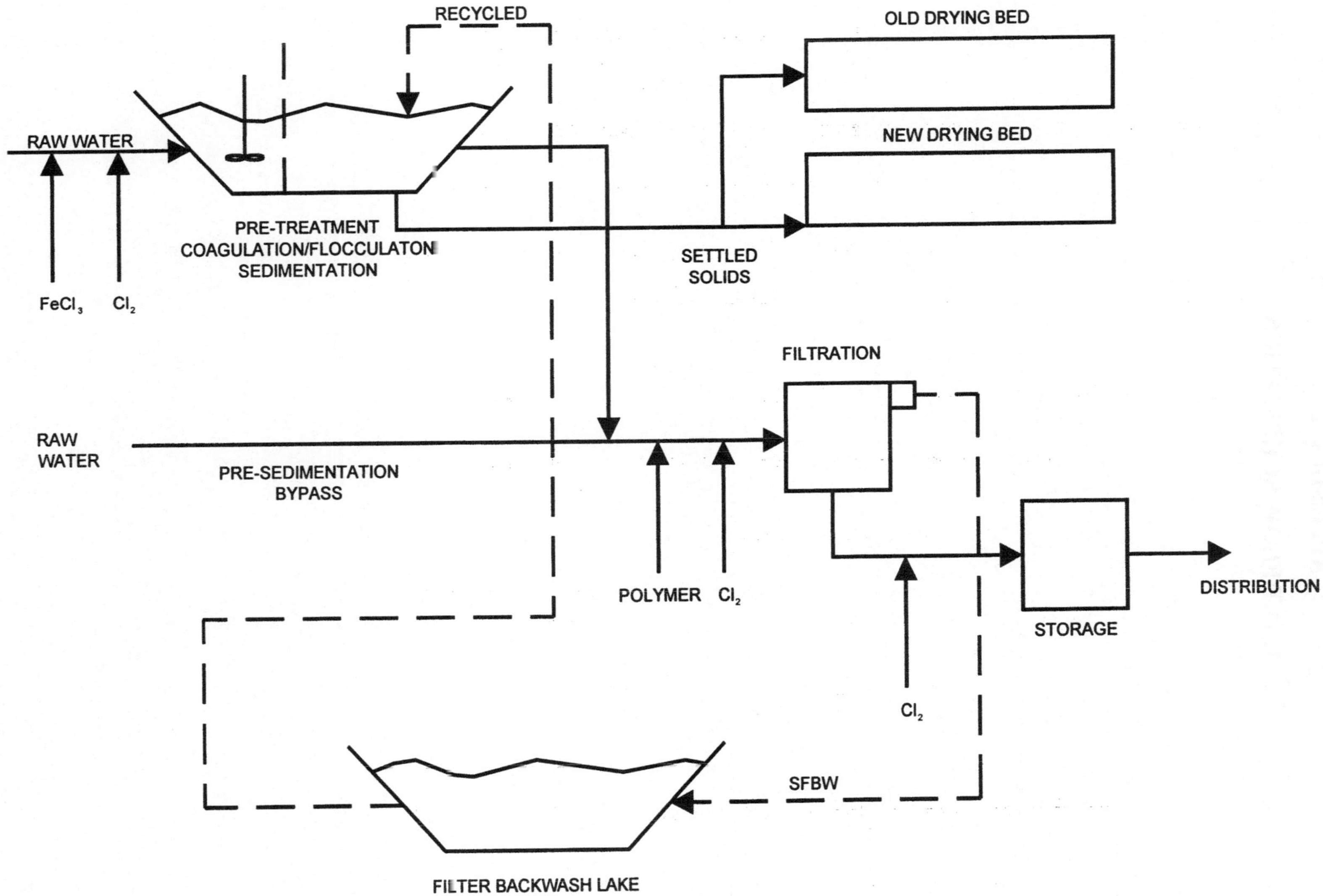

Figure A.1 City of Billings, MT process schematic

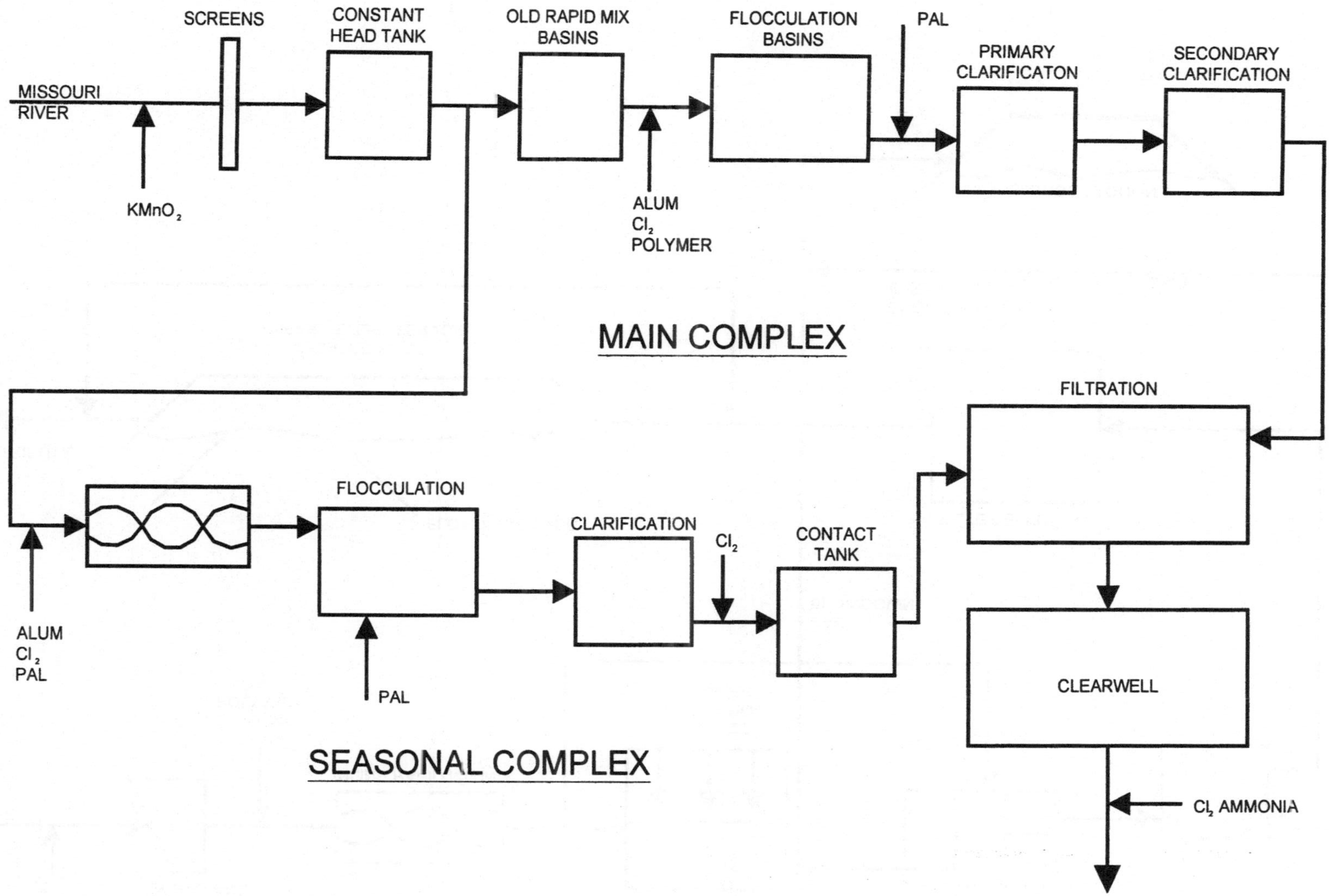

Figure A.2 City of Great Falls, MT process schematic

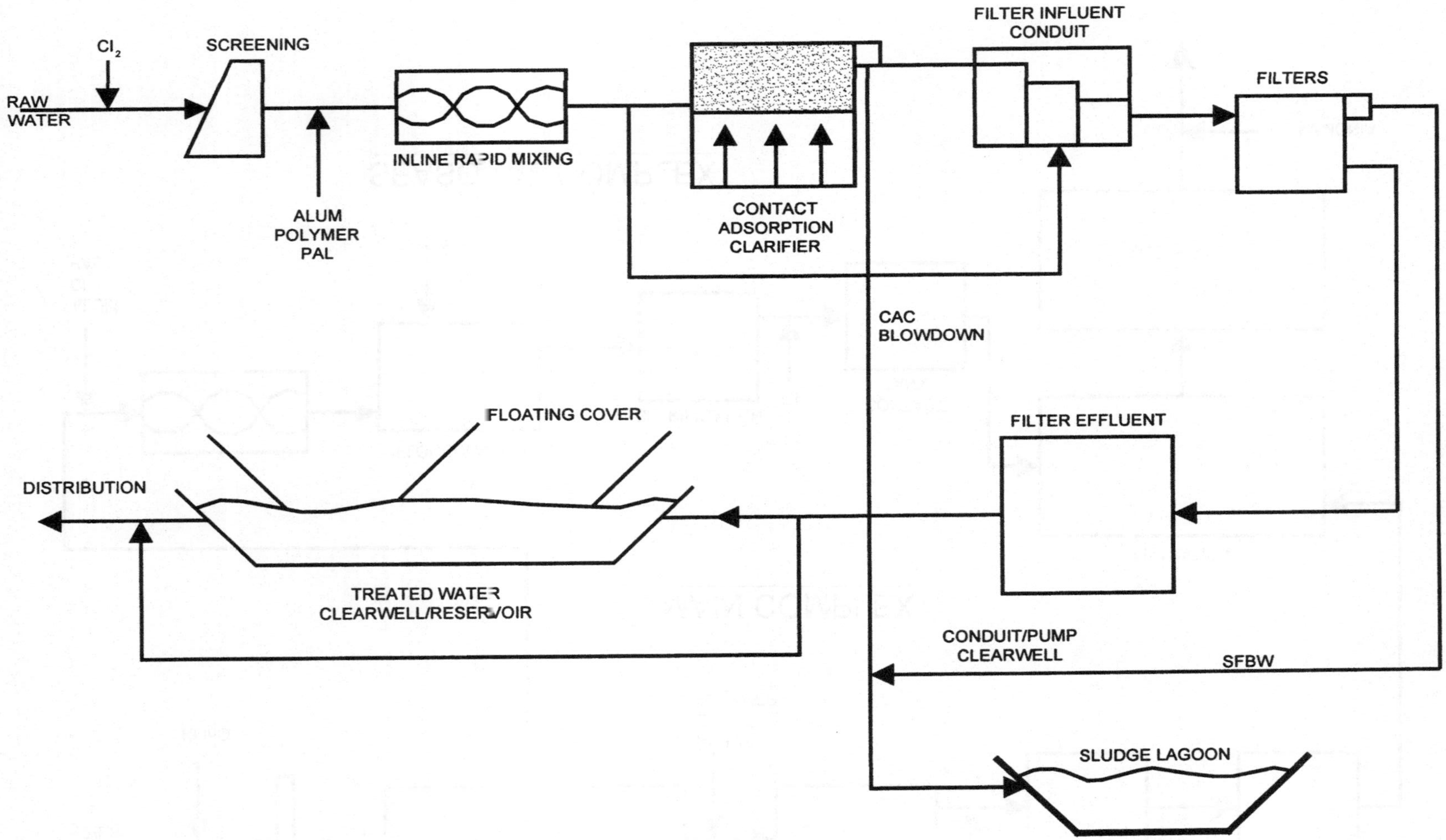

Figure A.3 City of Helena, MT process schematic

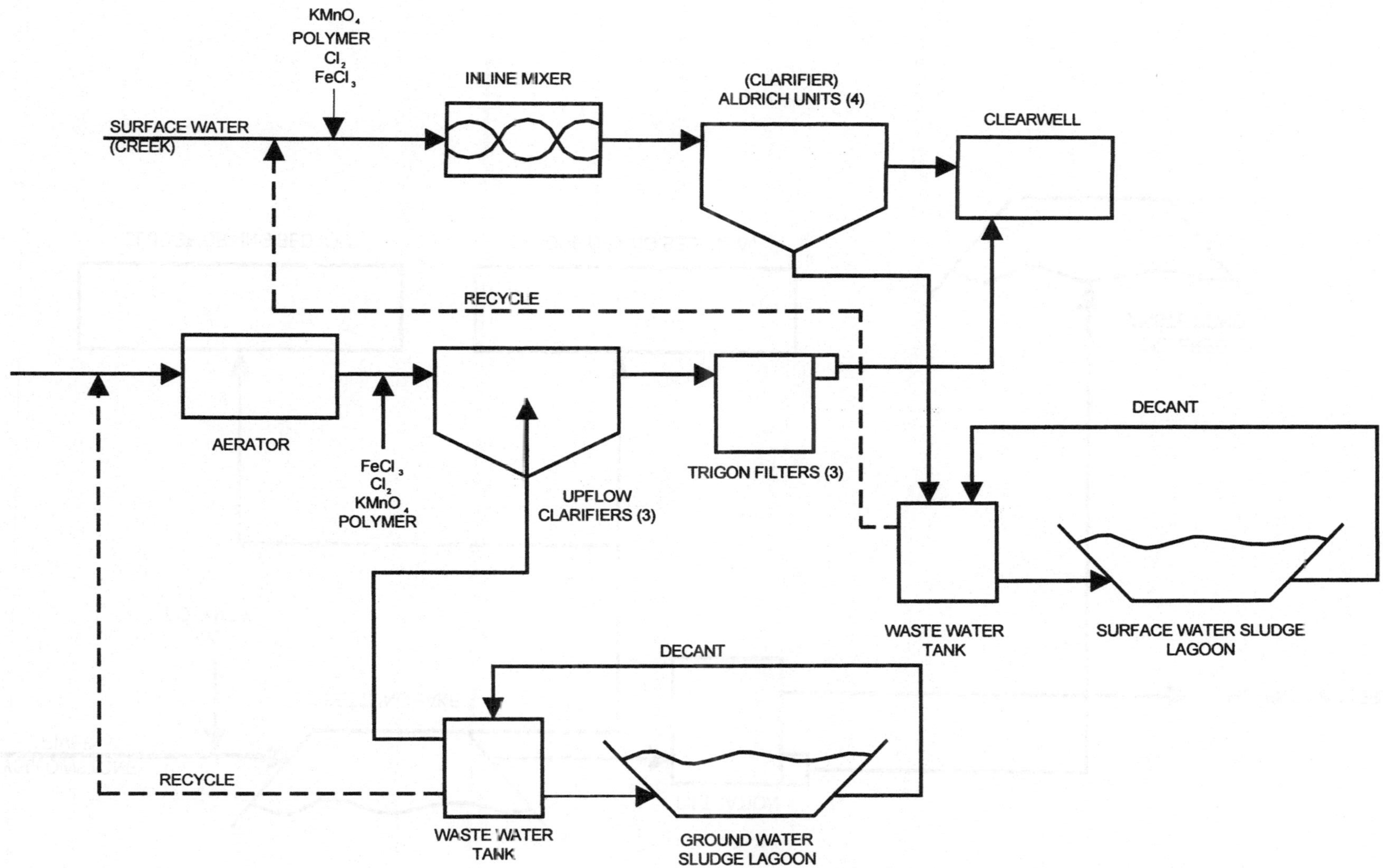

Figure A.4 Indiana-American, Kokoma, IN process schematic

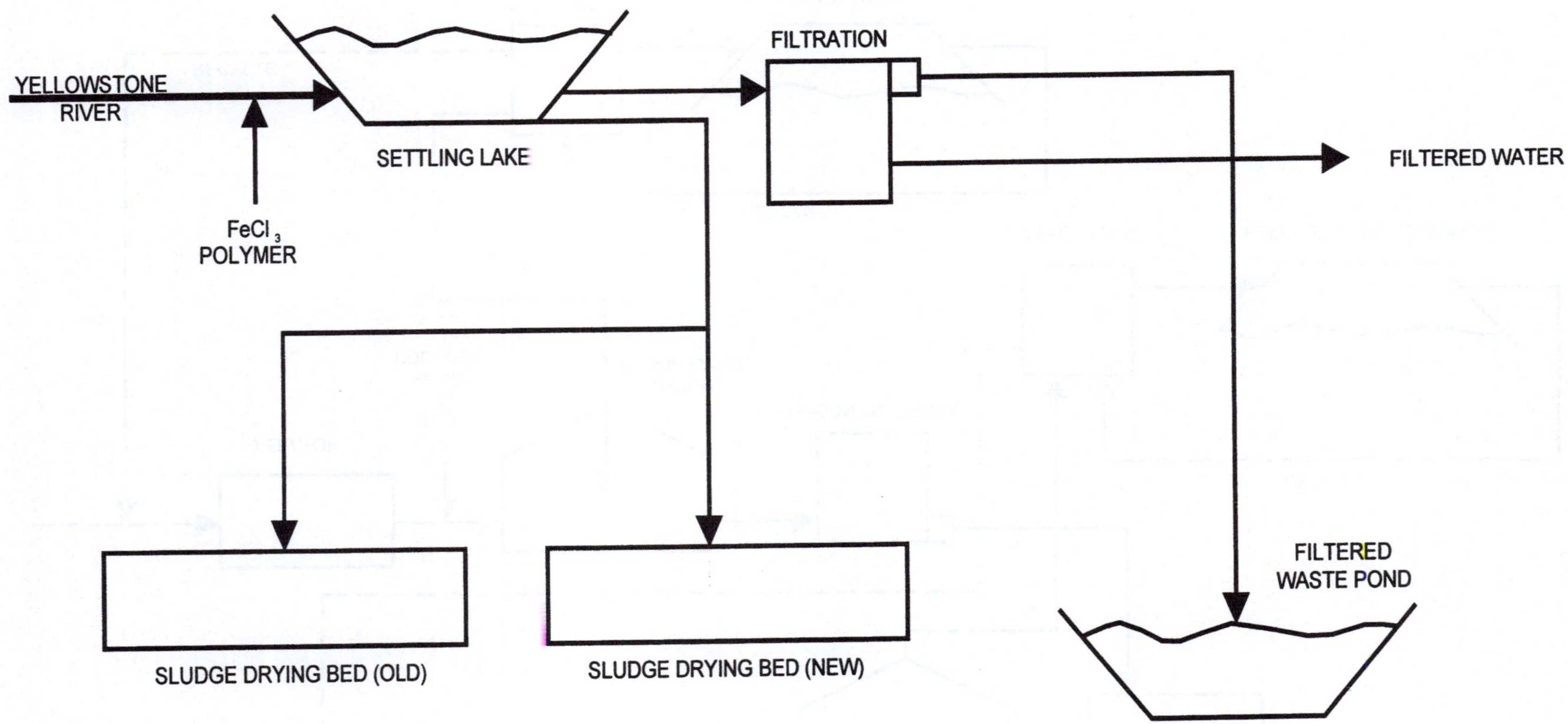

Figure A.5 Lockwood, MT process schematic

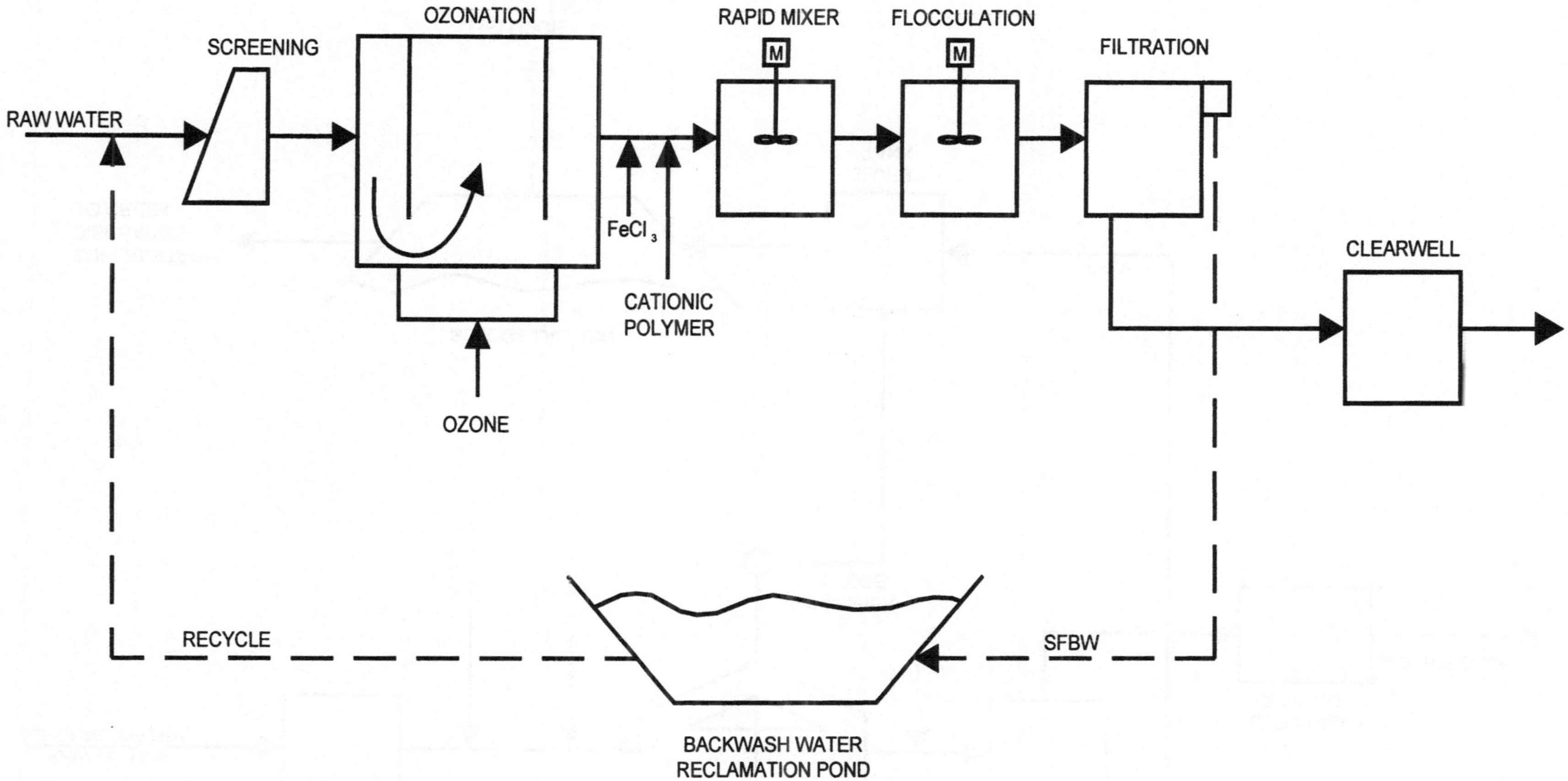

Figure A.6 Los Angeles, CA aquaduct process schematic

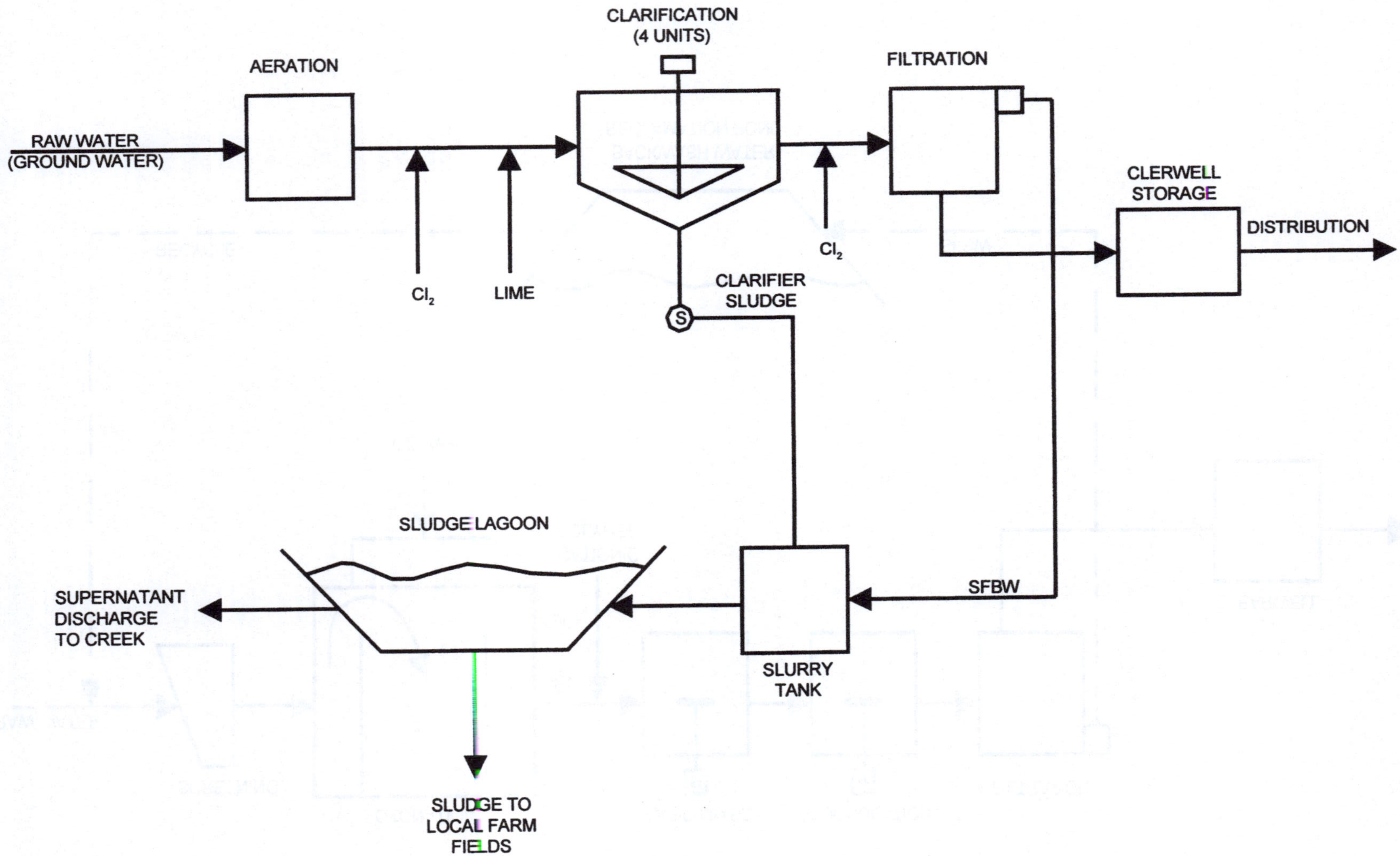

Figure A.7 New Iberia, LA process schematic

APPENDIX B

SIMPLE CALCULATIONS FOR COMPARISON OF LAGOON ARSENIC LEACHING WITH TOXICITY TEST LEACHING

Equations:

1. Lagoon simulation

   - Determination of arsenic released per weight of dry solids in lagoon:

$$\text{Arsenic release (mg/kg)} = \frac{\text{Lagoon As release (mg / L)} * \text{Lagoon volume (L)}}{\text{Lagoon solids dry weight (kg)}}$$

2. Toxicity test (TCLP or Ca WET)

   - Determination of arsenic released per kg of dry solids used for analysis:

$$\text{Arsenic release (mg/kg)} = \frac{\text{Toxicity test As conc (mg / L)} * \text{Extraction volume (L)}}{\text{weight of dry solids analyzed (kg)}}$$

3. Ratio of toxicity test arsenic leaching (mg/kg) and lagoon arsenic leaching (mg/kg):

$$\text{Ratio (mg/mg)} = \frac{\text{Toxicity test As release (mg/kg)}}{\text{measured lagoon As release (mg/kg)}}$$

Sample Calculation:

Great Falls Example: Lagoon and TCLP arsenic release at 20 percent solids

1. Lagoon total arsenic released (mg/kg) = $\frac{0.55 \text{ mg/L} * 1.89 \text{ (L)}}{0.057 \text{ kg}} = 18.3 \text{mg / kg}$

2 TCLP arsenic release per kg dry solids (mg/kg) = $\frac{0.148 \text{ mg/L} * 0.3 \text{ (L)}}{0.003 \text{ kg}} = 41.8 \text{mg / kg}$

3. Ratio of TCLP arsenic to lagoon arsenic release (mg/mg) = $\frac{41.8 \text{ mg/kg}}{18.3 \text{ mg/kg}} = 2.3$

Table B.1

Summary of lagoon arsenic release and TCLP test arsenic release (20-percent solids concentration)

Utility name	Lagoon solids Conc. (percent)	Lagoon solids dry wt. (kg)	Lagoon supernatant As Conc. (mg/L)	Lagoon total As released per wt. of dry solids (mg/kg)	20 percent solids TCLP As Conc. (mg/L)	20 percent solids wt. of dry solids for TCLP (kg)	20 percent solids TCLP As release per kg dry solids (mg/kg)	Ratio of TCLP As (@ 20 percent) to lagoon As release
Great Falls, MT	3.0	0.057	0.550	18.33	0.418	0.003	41.8	2.3
Helena, MT	3.6	0.069	0.660	18.13	0.039	0.003	3.9	0.2
Indiana American	1.9	0.035	0.100	5.38	0.016	0.003	1.6	0.3
Lockwood, MT	4.7	0.089	0.027	0.57	0.010	0.003	1.0	1.8
Los Angeles (typical)	4.0	0.076	0.870	21.64	0.008	0.003	0.8	0.0
Louisiana WC	4.1	0.077	0.005	0.12	0.006	0.003	0.06	4.9
Billings, MT	18.6	0.352	0.640	3.44	0.066	0.003	6.6	1.9

Note: TCLP extraction volume = 0.3L

Lagoon volume = 1.89L

Table B.2

Summary of lagoon arsenic release and TCLP test arsenic release (100-percent solids concentration)

Utility name	Lagoon solids conc. (percent)	Lagoon solids dry wt. (kg)	Lagoon supernatant As conc. (mg/L)	Lagoon total As released per wt. of dry solids (mg/kg)	100 percent solids TCLP As conc. (mg/L)	100 percent wt. of dry solids for TCLP (kg)	100 percent solids TCLP As release per kg dry solids (mg/kg)	Ratio of TCLP As (@100 percent) to lagoon As release
Great Falls, MT	3.0	0.057	0.550	18.33	0.048	0.015	1.0	0.1
Helena, MT	3.6	0.069	0.660	18.13	0.130	0.015	2.6	0.1
Indiana American	1.9	0.035	0.100	5.38	0.011	0.015	0.2	0.0
Lockwood, MT	4.7	0.089	0.027	0.57	0.009	0.015	0.2	0.3
Los Angeles (typical)	4.0	0.076	0.870	21.64	0.079	0.015	1.6	0.1
Louisiana WC	4.1	0.077	0.005	0.12	0.095	0.015	1.9	15.4
Billings, MT	18.6	0.352	0.640	3.44	0.010	0.015	0.2	0.1

Note: TCLP extraction volume = 0.3L

Lagoon volume = 1.89L

Table B.3

Summary of lagoon arsenic release and Ca WET test arsenic release (20-percent solids concentration)

Utility name	Lagoon solids Conc. (percent)	Lagoon solids dry wt. (kg)	Lagoon supernatant As Conc. (mg/L)	Lagoon total As released per wt. of dry solids (mg/kg)	20 percent solids Ca WET As Conc. (mg/L)	20 percent solids wt. of dry solids for WET (kg)	20 percent solids WET As release per kg dry solids (mg/kg)	Ratio of WET As (@20 percent) to lagoon As release
Great Falls, MT	3.0	0.057	0.550	18.33	7.367	0.003	368.4	20.1
Helena, MT	3.6	0.069	0.660	18.13	2.730	0.003	136.5	7.5
Indiana American	1.9	0.035	0.100	5.38	74.666	0.003	3733.3	694.4
Lockwood, MT	4.7	0.089	0.027	0.57	4.123	0.003	206.2	361.1
Los Angeles (typical)	4.0	0.076	0.870	21.64	3.570	0.0063	178.5	8.2
Louisiana WC	4.1	0.077	0.005	0.12	0.152	0.003	7.6	61.5
Billings, MT	18.6	0.352	0.640	3.44	10.718	0.003	535.9	155.7

Note: California WET extraction volume = 0.15L

Lagoon volume = 1.89L

Table B.4

Summary of lagoon arsenic release and Ca WET test arsenic release (100-percent solids concentration)

Utility name	Lagoon solids Conc. (percent)	Lagoon solids dry wt. (kg)	Lagoon supernatant As Conc. (mg/L)	Lagoon total As released per wt. of dry solids (mg/kg)	100 percent solids Ca WET As Conc. (mg/L)	100 percent solids wt. of dry solids for WET (kg)	100 percent solids WET As release per kg dry solids (mg/kg)	Ratio of WET As (@ 100 percent) to lagoon As release
Great Falls, MT	3.0	0.057	0.550	18.33	6.111	0.015	61.1	3.3
Helena, MT	3.6	0.069	0.660	18.13	5.608	0.015	56.1	3.1
Indiana American	1.9	0.035	0.100	5.38	10.232	0.015	102.3	19.0
Lockwood, MT	4.7	0.089	0.027	0.57	0.260	0.015	2.6	4.6
Los Angeles (typical)	4.0	0.076	0.870	21.64	15.000	0.015	150.0	6.9
Billings, MT	18.6	0.352	0.640	3.44	2.256	0.015	22.6	6.6

Note: California WET extraction volume = 0.15L

Lagoon volume = 1.89L

APPENDIX C

SAMPLE COLLECTION INSTRUCTION

Sample collection instructions:

1. Collect first grab sample of SFBW as it first spills into the backwash troughs at the start of a typical filter backwash using a sample collection device such as a wide-mouth sample bottle or bucket tied to a string, connected to a broom handle, etc.
2. Collect subsequent grab samples of equal volume from the same point (as the water flows into the troughs) every minute for the duration of the backwash
3. Combine grab samples in one container to make a representative composite SFBW sample for analysis

Please note that SFBW sample should be collected during a normal backwash after a filter has been on line for a typical number of hours. Please do not collect a grab sample after a filter has been on line for only a short time because such a sample would be too "clean" and not representative.

REFERENCES

Amy, G., and M. Edwards. 1999. Residuals Generation, Handling and Disposal. In Arsenic Treatment Options and Residuals Handling Issues. Draft Final Report.

Bothe, Jr., James V. and Paul W. Brown. 1999. Arsenic Immobilization by Calcium Arsenate Formation. *Environ. Sci. Technol.*, 33:3806.

Chen, H. 2001. Ph.D. Dissertation in preparation. University of Colorado, Boulder.

Cornwell, D.A., C. Vandermeyden, G. Dillow, and M.Wang. 1992. Final Report on Landfilling of Water Treatment Plant Coagulant Sludges. Denver, CO.: AwwaRF

Cornwell, D.A., and H.M.M. Koppers, eds. 1990. *Slib, Schlamm, Sludge.* Denver, CO.: AwwaRF and KIWA.

Cornwell, D.A., M. MacPhee, and R. Brown. 2002. *Occurrence of, and Problems Associated with, Trace Contaminants in Water Treatment Chemicals.* Denver, CO: AwwaRF, forthcoming.

Dzombak, David, A. Morel, and M.M. Francois. 1990. *Surface Complexation Modeling: Hydrous Ferric Oxide.* New York: John Wiley and Sons, Inc.

Frey, Michelle, and M. Edwards. 1997. Surveying Arsenic Occurrence. *Jour. AWWA*, 89:2:107.

Smith, Sheryl D. and Marc A. Edwards. 2001. Role of Silica and Calcium in Arsenic Uptake by Amorphous and Granular Media Oxides. Submitted to *Jour. AWWA*.

APHA, AWWA, and WEF (American Public Health Association, American Water Works Association, and Water Environment Federation). 1998. *Standard Methods for the Examination of Water and Wastewater*. 20th ed. Washington, D.C.: APHA.

U.S. Environment Protection Agency. 2000. Regulations on the Disposal of Arsenic Residuals from Drinking Water Treatment Plants. May 2000. EPA/600/R-00/025. Office of Research and Development, Washington, D.C.

U.S. Environment Protection Agency. 2001. Treatment of Arsenic Residuals from Drinking Water Removal Processes. June 2001. EPA/600/R-01/033. Office of Research and Development, Washington, D.C.

Villalobos, Mario, and James O. Leckie. 2001. Surface Complexation Modeling and FTIR Study of Carbonate Adsorption to Goethite. *Journal of Colloid and Interface Science,* 235:15.

ABBREVIATIONS

AA	activated alumina
Al	aluminum
As	arsenic
AsO_4^{3-}	arsenate
AZ	Arizona
BDL	below detection limit
BOM	biodegradable organic matter
BPJ	best professional judgment
CA	California
Ca	calcium
Ca WET	California waste extraction test
CFR	code of federal regulations
CMF	coagulation microfiltration
CO	Colorado
CO_2	carbon dioxide
CO_3^{2-}	carbonate
CWA	Clean Water Act
DEQ	Department of Environmental Quality
DHES	Department Health Environmental Services
DO	dissolved oxygen
DPHE	Department Public Health and Environment
EE&T	Environmental Engineering & Technology, Inc.
EPA	Environmental Protection Agency
FAV	final acute value
FBRR	Filter Backwash Recycle Rule
Fe	iron
$FeCl_3$	ferric chloride
FeH_2AsO_4	ferrous arsenate

FeOCOOH	ferrous oxide formate
FeOH	ferric
gal	gallons
H^+	hydrogen ion
HNO_3	hydrochloric acid
H_2O	water
ICP-AES	inductively coupled plasma atomic emission spectroscopy
IDEM	Indiana Department of Environmental Management
IMD	Injection and Mining Division
IN	Indiana
IX	ion exchange
KDHE	Kansas Department of Environmental Health and System
KS	Kansas
LA	Louisiana
LADES	Louisiana Department of Environmental Quality
LADWP	Los Angeles Department of Water and Power
MCL	maximum contaminant level
ME	Maine
MG	million gallons
Mg	magnesium
$Mg(OH)_2$	magnesium hydroxide
mg/L	milligrams per liter
mg/kg	milligrams per kilograms
MN	Minnesota
Mn	manganese
MPCA	Minnesota pollution control agency
MPDES	Montana pollutant discharge elimination system
MT	Montana
NaOH	sodium hydroxide
ND	not detected

NE	Nebraska
NF	nonofiltration
NH	New Hampshire
NHDES	New Hampshire Department of Environmental Services
NM	New Mexico
NPDES	National pollutant discharge elimination system
NV	Nevada
OFR	overflow rate
OWQ	Office of Water Quality
PA	Pennsylvania
POTW	publicly owned treatment works
ppb	parts per billion
RCRA	Resource Conservation and Recovery Act
RO	reverse osmosis
SDS	state disposal system
SDWA	Safe Drinking Water Act
SFBW	spent filter backwash water
TAC	Texas administrative code
TBLL	technically based local limit
TCLP	toxicity characteristic leaching procedure
TMDL	total maximum daily load
TNRCC	Texas National Resources Conservation Commission
TPDES	Texas pollutant discharge elimination system
TSS	total suspended solids
TX	Texas
μg/L	micrograms per liter
UIC	underground injection control
UPDES	Utah pollutant discharge elimination system
UT	Utah
WDNR	Wisconsin Department of National Resource

WI	Wisconsin
WPDES	Wisconsin pollutant discharge elimination system
WQBELS	water quality based effluent limits
WTP	water treatment plant
WWTP	wastewater treatment plant

IWA Publishing's authorised EU representative for General Product Safety Regulations is Diane D'Arras, 15 rue Duret, 75116 Paris, France, e-mail: safety@iwap.co.uk.

Printed and bound by CPI Group (UK) Ltd, Croydon, CR0 4YY
17/09/2026
02218810-0002